STRUCTURE AND AGENT IN THE SCIENTIFIC DIPLOMACY OF CLIMATE CHANGE

ADVANCES IN GLOBAL CHANGE RESEARCH

VOLUME 5

The titles in this series are listed at the end of this volume.

STRUCTURE AND AGENT IN THE SCIENTIFIC DIPLOMACY OF CLIMATE CHANGE

An Empirical Case Study of Science-Policy Interaction in the Intergovernmental Panel on Climate Change

by

Tora Skodvin
Center for International Climate and Environmental Research - Oslo (CICERO),
University of Oslo,
Blindern, Oslo, Norway

KLUWER ACADEMIC PUBLISHERS
DORDRECHT / BOSTON / LONDON

Library of Congress Cataloging-in-Publication Data

Skodvin, Tora.
Structure and agent in the scientific diplomacy of climate change : an empirical case study of science-policy interaction in the Intergovernmental Panel on Climate Change / by Tora Skodvin.
p. cm. -- (Advances in global change research ; v. 5)
Includes bibliographical references and index.
ISBN 978-90-481-5588-0
1. Climatic changes--Government policy. 2. Intergovernmental Panel on Climate Change. 3. Science and state. I. Title. II. Series.

QC981.8.C5 S46 2000
363.738'745--dc21
ISBN 978-90-481-5588-0 00-064715

Published by Kluwer Academic Publishers,
P.O. Box 17, 3300 AA Dordrecht,,The Netherlands.

Sold and distributed in North, Central and South America
by Kluwer Academic Publishers,
101 Philip Drive, Norwell, MA 02061, U.S.A.

In all other countries, sold and distributed
by Kluwer Academic Publishers,
P.O. Box 322, 3300 AH Dordrecht, The Netherlands.

Printed on acid-free paper

ISBN 978-90-481-5588-0 ISBN 978-0-306-48168-0 (eBook)
DOI 10.1007/978-0-306-48168-0

Softcover reprint of the hardcover 1st edition 2000

Contents

ACRONYMS

AGBM:	Ad hoc Group on the Berlin Mandate
AGGG:	Advisory Group on Greenhouse Gases
AOSIS:	Alliance of Small Island States
BATNA:	Best Alternative To a Negotiated Agreement
CFC:	chlorofluorocarbons
CO:	carbon monoxide
COP:	Conference of the Parties to the UNFCCC
EC:	European Community
EPA:	Environmental Protection Agency (USA)
GCC:	Global Climate Coalition
GCM:	General Circulation Model
GHG:	greenhouse gases
GWP:	global warming potential
HCFC:	hydrochlorofluorocarbons
HFC:	hydrofluorocarbons
ICSU:	International Council of Scientific Unions
IGO:	intergovernmental organisation
INC/FCCC:	Intergovernmental Negotiating Committee for a Framework Convention on Climate Change
INC:	See INC/FCCC
IPCC:	Intergovernmental Panel on Climate Change
LA:	lead author
LRTAP:	long-range transboundary air pollution
NGO:	non-governmental organisation
ODP:	ozone depleting potential
OECD:	Organisation for Economic Co-operation and Development
SBSTA:	Subsidiary Body for Scientific and Technical Advice
SPM:	Summary for Policymakers
UNCED:	United Nations Conference on Environment and Development
UNEP:	United Nations Environment Programme
UNFCCC:	United Nations Framework Convention on Climate Change
WCP:	World Climate Programme

WG:	working group (of the IPCC)
WGI:	Working Group one
WGII:	Working Group two
WGIII:	Working Group three
WMO:	World Meteorological Organization

Acknowledgements

This book is a slightly revised version of my PhD thesis in political science at the University of Oslo. I am indebted to Prof. Arild Underdal at Department of Political Science, University of Oslo, for his invaluable supervision of my work with this project. I have benefited and learned enormously from his systematic approach, his wise comments and insights, and his extraordinary ability to provide guidance, inspiration and stimulation. Without his conscientious supervision my task would have been so much more difficult.

I am very grateful to Karen Litfin (University of Washington), Albert Weale (University of Essex) and Raino Malnes (University of Oslo) for the thoroughness with which they commented and evaluated my thesis.

I doubt whether this analysis could have been carried out without the ability to participate at IPCC meetings. My participation in these meetings was made possible by the Norwegian Ministry of Environment, who also provided some travelling funds, and the Technical Support Unit of Working Group I. I am particularly grateful to Håvard Thoresen and Øyvind Christophersen at the Norwegian Ministry of Environment and Judy Lakeman at the Technical Support Unit of Working Group I who made my participation at IPCC meetings possible.

I am grateful to Sonja Boehmer-Christiansen (University of Hull), who gave me access to her personal file on the IPCC, which generated very valuable information for my empirical analysis. I have also benefited greatly from discussions with Simon Shackley (University of Manchester), Nils Roll-Hansen (University of Oslo), Jon Birger Skjærseth, Steinar Andresen and Jørgen Wettestad (all at the Fridtjof Nansen Institute in Oslo).

An important source of information has been interviews with a number of people (listed in the appendix). I gratefully appreciate that they took the time to answer my questions despite their tight time schedules.

I am indebted to Center for International Climate and Environmental Research – Oslo (CICERO) for giving me the flexibility, time and funds to carry out my analysis, particularly Helga Hernes (director until August 1996) and Knut H. Alfsen (director since January 1997). The interdisciplinary composition of CICERO's staff has provided me with a unique opportunity to explore new fields and perspectives, and my colleagues at CICERO have been a source of inspiration and learning. I am particularly grateful to Jan S. Fuglestvedt and Ivar S. A. Isaksen who have taught me (almost) all I know about atmospheric chemistry and the climate change problem as a field of research within natural science.

In the preparation of this manuscript for publication, Lynn P. Nygaard, who has done the language- and technical-editing of the manuscript, has made an invaluable contribution.

Last, but not least, I am indebted to my husband, Torgrim Eikeland, and our two daughters, Sofie and Agnes, without whose practical and moral support this project could not have been carried out.

The responsibility for errors and misinterpretations rests with the author.

The project was financed by the Norwegian Research Council, grant no. 101975/730.

Tora Skodvin
Oslo, 1 July, 1999

Chapter 1

Introduction

1.1 Structure and Agent in Science–Policy Interaction

Research input constitutes a key component in the development of effective international environmental regimes. As it appears in journals and other scientific publications, however, scientific knowledge is not readily available for policy-makers to apply in a particular policymaking context. While relevant, scientific knowledge usually does not explicitly address the particular problems policy-makers struggle with. Scientific knowledge is also produced within a context – within structures of knowledge and theories that constitute the basis for the knowledge generated – which implies that scientific knowledge is characterised by a certain context dependency. Also, scientific knowledge is provided in a technical form that is not applicable in policymaking. Thus, for scientific knowledge to be applicable in policymaking, it needs to be interpreted and translated – transformed – into a form in which it may serve as a premise for policy choice.

In this process of transformation, the competence of both scientists and policy-makers is needed. Scientific competence is needed to ensure that the knowledge base provided is representative of state-of-the-art knowledge within relevant fields and disciplines. Policy competence is needed to ensure its relevance and applicability to the particular policy problem for which the input is sought and developed. Thus, scientific knowledge for policymaking is generated in processes of science–policy interaction.

Science–policy interaction is difficult and demanding because of its immanent tension between impartiality and disinterestedness on the one hand, and strategic behaviour and interest realisation on the other. This tension is generated by the interaction between two distinctively different systems of behaviour. While science (ideally) is conceived of as a truth-seeking endeavour – whose norms and guidelines for behaviour are directed towards the generation of "objective" and disinterested knowledge – politics constitutes a system for the generation of (collective) decisions, where behaviour is directed towards the realisation of (individual) rational interests in these decisions. In contrast to the ideal of impartiality characterising the scientific method, political behaviour is characterised by strategic reasoning where the instrumental utilisation – as well as manipulation and distortion – of knowledge may constitute central elements in political strategies whereby individual interests are sought realised. This tension is reinforced, moreover, by an image of the relationship between science and politics as one of opposite poles, where science is everything politics is not: pure, objective, subject to rational analytical reasoning and thus not hostage to manipulation tactics and coercive power – ingredients often associated with politics. While both theoretical analyses and experience show that the relationship between science and politics by far is as clear-cut as this image suggests, this image has a strong position in the public mind as well as among practising scientists and policy-makers themselves. Thus, any interactive dialogue between these two systems of behaviour takes place in the shadow of this image which suggests that the interaction itself implies a risk of political "contamination" of the scientific process and a serious loss of legitimacy.

Participants to processes of science–policy interaction thus face at least two difficult challenges in their endeavour. First, the ideal of scientific knowledge as disinterested and objective implies that it has a powerful force as legitimisation for policy arguments (Litfin, 1994). Thus, policymakers may engage in an effort to couch their policy arguments in scientific terms: Each new policy argument is introduced with a scientific justification and a corresponding delegitimisation of the arguments of their opponents. In this manner, processes of science–policy interaction may take on a vicious and highly counterproductive dynamic of legitimisation and delegitimisation (Litfin, 1994). This dynamic can also be expected to be reinforced by political conflict. Thus, the tension between impartiality and interest realisation inherent in science–policy interaction may give rise to a dynamic whereby the well-functioning and effectiveness of the process is at stake. Second, particularly as a function of the popular image of science–policy relations, processes of science–policy interaction balance a thin edge between integrity and isolation on the one hand, and relevance and applicability on the other. Once this "balance" tilts in the direction of the one or the other, either the policy relevance and applicability of the knowledge base provided is threatened, or the scientific legitimacy and authority of the process is jeopardised.

This book argues that processes of science–policy interaction take place within institutional structures that to a varying extent are capable of handling the obstacles and challenges associated with science–policy interaction. Thus, it is assumed that the institutional structures within which the science–policy dialogue takes place may have an independent impact on the extent to which the dilemmas associated with science–policy interaction are resolved, and may hence have an independent impact on the effectiveness of the process. To the extent that institutions capable of handling these challenges may be developed by conscious design, institutional design may represent a potential instrument whereby the effectiveness of processes of science–policy interaction may be enhanced. The first objective of this analysis is to explore and investigate this proposition.

The instrumental potential of institutional structures arises from the socially constructed nature of institutions: Institutions are social constructions and may as such – in principle – be utilised as instruments to improve the functioning and effectiveness of the process. While institutions are socially constructed, however, they are not necessarily the result of a conscious design: They may, for instance, emerge as a result of natural evolution. Thus, the instrumental potential of institutions is not constant. It may vary with context and the type of institution in question. This implies that the capacity of an institutional structure to enhance the effectiveness of processes of science–policy interaction may be a matter of degrees, and that the level of effectiveness may vary strongly even between cases with a comparable design. One factor which may contribute to the variable effect of institutional design is actor behaviour.

While the capacity of an institutional structure to improve institutional performance is based on its impact on the behaviour of agents operating within these structures, the behaviour of agents may also affect the way an institutional structure works. Actors are, evidently, not marionettes completely controlled by institutional arrangements, and there is, therefore, no "one-to-one" relationship between institutional design and actor behaviour. One form of actor behaviour, however, may be particularly important for the way institutional arrangements work: leadership performance. The second objective of this analysis is, therefore, to explore the role and impact of leadership in processes of science–policy interaction.

The provision of leadership depends on the ability of individual agents to convert a leadership potential into actual leadership. The provision of leadership, however, is also determined by the extent to which the institutional framework provides the necessary mandate and flexibility for this mode of behaviour. In the same manner that the impact and effect of institutional arrangements may depend upon the provision of leadership performance, a successful provision of leadership may depend upon institutional arrangements. The relationship between these two factors is thus assumed to be characterised by interdependency and "causal complexity" as discussed by Ragin (1987). Causal complexity refers to a situation where it is the combination of conditions that produces a change – which is different from saying that each variable in itself produces a change in another variable:

> The basic idea is that a phenomenon or change emerges from the *intersection* of appropriate preconditions – the right ingredients for change. In the absence of any one of the essential ingredients, the phenomenon – or the change – does not emerge. This conjunctural or combinatorial nature is a key feature of causal complexity. (Ragin, 1987: 25, emphasis added).

This analysis thus has three objectives: First, it sets out to investigate the extent to which and the manner in which institutional design can be utilised as an instrument to enhance the effectiveness of processes of science–policy interaction. Second, it investigates the impact of leadership performance on the functioning and effectiveness of processes of science–policy interaction. Finally, it seeks to explore the interdependent nature of the relationship between institutional design and actor behaviour in the form of leadership performance. In this way, the analysis investigates the impact and relationship of structures and agents in science–policy interaction.

This analytical framework is employed in an empirical case study of the Intergovernmental Panel on Climate Change (IPCC) within which the science–policy dialogue associated with the negotiation process on measures to abate a human-induced climate change has taken place. Together with issues such as the loss of biodiversity and the depletion of the ozone layer, the problem of a human-induced climate change belongs in the category that has been referred to as "the new generation of environmental problems" (Hajer, 1995). These phenomena are all symptoms of environmental degradation that remain invisible to the lay public decades after they have taken effect. Scientific knowledge is thus called for not only in the design of policies that are effective in terms of solving the problems for which they were designed, but also in the identification of the problem itself. Climate change is characterised by a high level of political conflict. It is scientifically complex. And it is beset with uncertainty. Climate change, therefore, is particularly illustrative of the nature and dynamics of science–policy interaction, and thus represents a rich source of data for this type of analysis.

1.2 The Scientific Diplomacy of Climate Change

The signing of the Kyoto Protocol of December 1997 marks a preliminary end to a six-year process of intense negotiations over whether the international political community should reduce their emissions of greenhouse gases (GHGs) in response to scientists' warnings that the sustained increase in atmospheric GHG concentrations caused by human activities would alter the global climate. In 1992, negotiations led to the signing of the United Nations Framework Convention on Climate Change in which the problem of a human-induced climate change is acknowledged, but which does not hold any binding commitments to reduce emissions. Following the entering into force of the climate convention in 1994, negotiations on a protocol started in earnest in 1995. The negotiation process (temporarily) culminated with the Kyoto Protocol, agreed upon by the Conference of the Parties to the convention (COP) in December 1997. The Kyoto Protocol has not yet been ratified, and has thus not entered into force.

The Kyoto agreement implies a legally binding commitment by industrialised countries to a total GHG emissions reduction by 2012 of 5.2% as compared to emissions levels in 1990. The agreement entails a set of differentiated control targets: While the United States and the European Union have committed themselves to 7% and 8% emissions reductions (respectively), other countries are permitted to increase their greenhouse gas emissions by up to 10% (Norway: +1%, Australia: +8% and Iceland: +10%). While this agreement is likely only to have a marginal impact on the accumulation of greenhouse gases in the atmosphere (Bolin, 1998), it represents an important first step towards a political strategy to abate the problem of a human-induced climate change. Above all, the Kyoto agreement is a strong indication of the extent to which a human-induced climate change has been accepted by the international political community as a political problem whose solution can only be found in concerted action.

This policy response is perhaps first and foremost the result of an effective negotiation process where conflicting interests and national concerns have been successfully integrated in the adoption of GHG reduction targets, timetables and implementation mechanisms that are acceptable to all parties. In this book, however, it is argued that this policy response is also the result of a relatively effective process of science–policy interaction through which a problem diagnosis has been developed – a problem diagnosis that is acceptable to all "stakeholders" (scientists as well as policymakers) and defines the problem of a human-induced climate change in scientific as well as political terms. This process has taken place within the framework of the Intergovernmental Panel on Climate Change (IPCC).

The IPCC was established in 1988 with a mandate to provide assessment reports of state-of-the-art knowledge on the various aspects of a possible human-induced climate change. It is made up of three Working Groups (WGs) that have assessed the science of climate change (WGI), possible impacts of climate change (WGII), and socio-economic and other cross-cutting issues (WGIII). The formulation of the task assigned to WGs II and III and the division of labour between these two WGs has varied somewhat during the course of the process. The IPCC is organised under the auspices of the World Meteorological Organisation (WMO) and the United Nations Environment Programme (UNEP). Thus, the IPCC process is unique in international relations for its institutional design: One of its main characteristics is that it has a scientific mandate while being organised as an intergovernmental body within a political institutional framework.

As a function of its intergovernmental status, the IPCC is characterised by broad participation of both scientists and policymakers and a relatively high level of formal integration between science and politics – much higher, it seems, than any other previous process of science–policy interaction in international environmental affairs. Levels of integration and participation patterns vary, however, within the three decision-making levels of the IPCC. With its "scientific core" at the bottom of the institution, where participation is dominated by scientists active in research, the level of integration between science and policy increases as we approach the upper decision making levels of the institution (the WG plenaries and the full panel plenary, which is the highest decision making level of the panel). Through this design, therefore, the institutional arrangements of the IPCC process serve both to separate and integrate science and politics. While this mode of operation has been criticised, this book argues that this institutional design has served to enhance the effectiveness of the process.

The analysis suggests that the capacity of the institutional arrangements of the IPCC to balance and combine scientific autonomy with science–policy integration seems to have contributed substantially to the extent to which policymakers have acknowledged the scientific authority of the knowledge base provided by the IPCC and accepted its conclusions as factually valid. In this case, policymakers' confidence in the research results communicated by scientists seems to be drawn from at least two main sources: first, the scholarly competence, integrity and independence of the scientists involved in the process, and second, the adversarial scrutiny of the knowledge base by actors and parties representing conflicting interests in the policy area. In an area as conflict-prone as climate change policies, the latter mechanism seems at least as crucial as the first.

It is important to note, however, that the analysis also suggests that the risk of politicisation and potentially paralysing conflicts inherent in an institutional design directed towards the adversarial scrutiny of the knowledge base by parties representing conflicting interests in the policy area has been counterbalanced by the leadership provided by key actors in the process. The analysis suggests that had important leadership functions not been provided in the process, the institutional design of the IPCC process could have contributed to an entirely different (opposite) outcome, and thus confirms the assumed contingency characterising the relationship between institutional structure and actor behaviour. In the IPCC process, leadership functions directed towards the development of the knowledge base, its transformation into decision premises and the provision of procedural management in this process, have enabled the IPCC to submit their reports to an intense scientific and political scrutiny without ending up in negotiation deadlock and impasse. It has also served to the provision of a knowledge base and a problem diagnosis which has acquired a high level of acceptance – within the scientific as well as the political communities – in an area characterised by significant scientific uncertainty.

1.3 Structure of the book

The analysis is conducted in two main parts. The first part develops the theoretical framework for the study. One aim of the analysis is to investigate ways in which the effectiveness of processes of science–policy interaction may be enhanced. In all analyses of international environmental management, the notion of effectiveness is difficult both to define and to measure. This is perhaps even more so the case with regard to processes of science–policy interaction. The notion of effectiveness is discussed in chapter 2. Usually, science–policy interaction is discussed in terms of the extent to which scientific knowledge has shaped, and is reflected in, the policy decision made. What may be referred to as a "traditional" definition of effectiveness is thus related to the realisation of the official purpose of processes of science–policy interaction: to generate scientific knowledge that is used as a premise for the policy decisions made. In chapter 2, an additional, and more fine-tuned definition of effectiveness is provided, according to which the extent to which policymakers have accepted the factual validity and scientific authority of the knowledge base provided is also considered to be an indication of effectiveness. Policymakers' acceptance of the knowledge base may be seen as a necessary condition for policy action in a given area of environmental management since it is difficult to imagine that policymakers act upon scientific knowledge whose factual validity they do not acknowledge. They may, however, "act upon" this knowledge for other reasons.

Chapter 3, explores the science–policy nexus. The point of departure for the analysis in this chapter is the nature and dynamics that characterise science and politics in their pure forms. The analysis of the nature and dynamics that characterise the process whereby scientific knowledge is produced draws upon literature from the philosophy and sociology of science. The analysis of the nature and dynamics that characterise the process whereby politics produces decisions builds on negotiation theory. On this basis, the nature and dynamics that characterise processes in which these two distinct systems of behaviour interact are explored and the main obstacles and pitfalls associated with science–policy interaction are identified.

In chapter 4, the extent to which and the manner in which institutional design may represent an instrument to overcome obstacles to science–policy interaction is analysed. The analysis in this chapter explores the extent to which institutions are designable on the basis of organisation theory. The causal pathways through which institutional arrangements are assumed to affect the outcome of processes of science–policy interaction are identified as the capacity of the institutional apparatus to serve four main functions: to maintain the scientific autonomy and integrity of scientists participating in the process; to provide a sufficient level of integration between science and politics; to ensure the geographic representativeness of scientific bodies; and, to provide mechanisms for conflict resolution.

Chapter 4 also investigates the potential role of leadership in a context of science–policy interaction. Leadership is explored on the basis of two main bodies of knowledge: the role of leadership in negotiations and the role of scientific elites in the generation of new knowledge within scientific communities. The analysis in this chapter identifies the different leadership functions that may be assumed to be in demand in processes of science–policy interaction. Three modes of leadership are identified: leadership functions directed towards the development of the knowledge base – particularly with regard to the scientific community's evaluation of the adequacy by which state-of-the-art knowledge is represented; leadership functions directed towards the transformation of the knowledge base into premises for decision-making; and, boundary–role leadership directed towards the provision of communicative links between scientists and policymakers as well as the development of institutional arrangements.

Finally, chapter 4 suggests two main alternative explanations to the level of policy acceptance of the knowledge base observed in this case. First, that the level of acceptance observed is a function of the configuration of political interests of parties (governments) in the policy area; and second, that the level of acceptance observed, is a function of the state of knowledge in this field of research.

In the second main part of the book, the empirical analysis is carried out. In chapter 5, the problem of a human-induced climate change is presented, and the effectiveness of the science–policy dialogue related to this problem is assessed in terms of the definition of effectiveness provided in chapter 2.

In chapter 6, the empirical analysis of the institutional design of the IPCC process, with an emphasis on WGI, is carried out. The chapter describes the institutional set-up of the IPCC, investigates the extent to which the institutional apparatus of the IPCC is capable of serving the four main functions identified in chapter 4, and discusses the extent to which this has had an impact on the effectiveness of the process.

In chapter 7, the empirical analysis of the impact of institutional arrangements in the IPCC process is supplemented by the analysis of different modes of leadership behaviour in the process, and the impact of this factor on its overall effectiveness.

The empirical analysis in chapters 6 and 7 is based on three main sources of data. First, one central source of data, is the official reports from IPCC meetings and sessions and unofficial correspondence and reports obtained from the archives of the Technical Support Unit of the IPCC's WGI.[1] The second, and most important source of data is personal participation as an observer to all IPCC plenary and WGI plenary meetings during the period from 1993–1995 and personal communication and informal interviews with the IPCC leadership and participants to the IPCC process.[2] Personal observation at IPCC meetings provided a special opportunity to acquire a unique source of data to trace, explore and investigate the mechanisms at work in this process. It also provided a very good basis for conducting interviews with central participants. This primary-source data has also, to the greatest extent possible, been related to and controlled against the third main source of data: available secondary literature on this subject.

In chapter 8, the robustness of these findings is controlled in light of a discussion of the extent to which the level of policy acceptance of the knowledge base observed in this case could have been produced by factors other than the institutional design of the process and the amount of leadership provided. The validity of the first alternative proposition is

[1] Some of the material is collected by myself. Some of it is obtained from the personal archive of Sonja Boehmer-Christiansen, but the original source of the material is the same.

[2] A list over interviewees and the IPCC meetings and sessions that have been observed is provided in the Appendix. The interviews were carried out as informal conversations and are referred to as "personal communication". A concern for discretion has implied that most references to the information generated in these communications is anonymised.

investigated in terms of the political malignancy of a human-induced climate change. The validity of the second alternative proposition is investigated both in terms of the conclusiveness of scientific knowledge in climate change and the IPCC's representation of scientific uncertainty.

In chapter 9, the findings of the analysis are summarised and discussed.

1.4 Generalisations

Generalisations on the basis of a qualitative case study should always be made with caution. External validity – whether the causal relationships present in the cases examined also can be expected to operate in a larger population of cases – depends upon the extent to which the cases studied have conditions that differ significantly from those of the larger population of cases (Mitchell and Bernauer, 1998). Internal validity is a precondition for external validity. In quantitative, large-N studies, ensuring internal validity also serves to ensure a significant external validity. In qualitative case studies, on the other hand, the ensuring of internal validity, for instance, by holding certain variables constant through a choice of cases, may simultaneously limit the range of cases to which one can validly generalise (Mitchell and Bernauer, 1998). In this regard, case studies provide a weaker basis for generalisations than quantitative research.

In the more specific context of international environmental affairs, one should also consider the possibility of path dependency between cases. While we have a tendency to treat each case as independent of other cases within a population of cases, they have often taken place in sequence, and may thus be linked by learning. Learning complicates the task of making valid generalisations.

The findings of this study must, therefore, be generalised to other cases only with great caution: This analysis investigates only one case of science–policy interaction and the scope of validity of the findings cannot be unequivocally determined. Processes of this kind are also context dependent, implying that the context within which different factors work also may have implications for how they work: That is, context-specific elements may represent conditions for the way different variables work that cannot be uncovered in a single-case analysis. On the other hand, the approach of the analysis is theoretically derived and should, therefore, in principle, be valid for similar processes of science–policy interaction associated with international environmental regime formation and policymaking. In all processes of science–policy interaction in international environmental regime formation and policymaking, the mechanisms described and analysed in this study can be expected to be at work, although they may take different forms. Similarly, the causal pathways through which independent and dependent variables are suggested to be linked – the four main functions outlined above – are also, in principle, valid for similar cases of science–policy interaction, although the specific institutional arrangements through which these functions are served may vary between cases. Thus, the analytical approach developed for the study is, in principle, applicable to all processes within the scope of analysis of this approach: processes of science–policy interaction associated with international environmental regime formation and policymaking.

The ability to generalise also depends upon the representativeness of the case which is subject to investigation, and hence the representativeness of the findings generated in the study. While the IPCC process represents a process of science–policy interaction *par excellence*, it is also a unique exemplar of its kind. There are no processes in international environmental affairs characterised by the level and extent of science–policy integration of the IPCC process. The historical development in processes of this kind indicates, however, that each new effort to design and develop an effective science–policy dialogue builds on the experience generated in previous efforts, and hence that they cannot be considered as entirely independent cases. Building on experience from previous efforts, each new case also adds a new and unique element which implies that it is not entirely similar to previous cases of the same phenomenon. In this progressive development, however, knowledge about the mechanisms at work in the IPCC experience should represent a valid input.

Chapter 2

Effectiveness in Processes of Science–Policy Interaction

2.1 Introduction

This analysis aims at investigating how the effectiveness of the science–policy dialogue associated with international environmental management may be enhanced. However, before discussing how effectiveness may be enhanced, it is important to establish just what is meant by "effectiveness" in this context and how it may be measured.

The effectiveness with which scientists have communicated scientific facts and findings to policymakers is often judged in terms of the extent to which scientific knowledge has constituted a premise for – and is reflected in – the policy decisions made. Policymakers' adoption and utilisation of scientific knowledge as premises for their decisions may be regarded as the official purpose of processes of science–policy interaction. The extent to which policymakers have acted upon scientific knowledge thus obviously constitutes an important measure and indication of effectiveness. It is, however, also a crude and very demanding measure: All encounters between scientists and policymakers that have not resulted in policy decisions clearly based on scientific knowledge would, according to this definition, be dismissed as ineffective. However, as discussed below, there are many other reasons why a policy does not reflect scientific knowledge. Thus, this definition is not sufficiently fine-tuned to capture the multifaceted nature of the science–policy dialogue. In this chapter, therefore, a definition of effectiveness is developed which not only employs policymakers' adoption of scientific knowledge as premise for their decisions as a measure of effectiveness, but which also may capture outcomes that may be characterised as effectiveness at a lower level – notably, the extent to which policymakers accept the scientific authority and factual validity of the findings and conclusions brought forward by scientific communities.

Effectiveness understood as the realisation of official purpose is discussed in section 2.2. Effectiveness judged in terms of policymakers' acceptance of factual validity is discussed in section 2.3. In section 2.4 a comprehensive definition of effectiveness is developed. Finally, in section 2.5, the notion of "consensual knowledge" is briefly discussed.

2.2 Effectiveness as Realisation of Official Purpose

In all matters of environmental and natural resource management, the link between science and policymaking is an important aspect. Moreover, with the introduction of the "new" global environmental problems on the international political agenda a new dimension has been added to science–policy relations. Issues such as loss of biodiversity, the depletion of the ozone layer and global warming are all issues that may remain invisible to the lay public decades after they have taken effect. Scientific input is thus needed not only in the process of designing solutions capable of solving the problem, but also in the process of identifying and detecting the problem in the first place (see, for instance, Hajer, 1995).

The official purpose of establishing channels for scientific input in processes of policymaking is to inform policymakers about basic cause-and-effect relationships characterising particular problems and thereby enable the development of policy decisions that are effective for dealing with the problems for which they were designed. Hence, the effectiveness of the science–policy dialogue can be seen as the extent to which this official purpose has been realised. In this conceptualisation, the science–policy dialogue is judged to be effective to the extent that policymakers have utilised and acted upon the scientific findings brought forward by the scientific community in the sense that scientific findings have constituted a premise for, and are reflected in, the decisions made.

This conceptualisation of effectiveness needs to be qualified in several respects. First, while the policy implications of a body of knowledge may be clearly spelled out in explicit scientific advice in some cases, in most cases policymakers are not provided with explicit recommendations and advice from the scientific community. During the last decades we have seen a trend where scientific input to policymaking increasingly takes the form of general assessments of state-of-the-art knowledge rather than explicit scientific advice providing clear guidance to behaviour. The development in science–policy relations in international environmental decision-making thus to some extent departs from the early experiences of science–policy interaction in international politics (around the 1940s and 50s), which mainly took place in relation to natural resource management, particularly the management of marine resources. This difference in the form in which scientific input is provided thus seems to be related to the nature of issue areas. In questions of resource management, explicit scientific advice for the prevention of over-exploitation (admittedly associated with a greater or lesser degree of scientific uncertainty and conclusiveness) is usually available. In such cases, effectiveness can be judged by comparing political decisions with the scientific advice provided. When political decisions correspond to the advice given, the process has been effective.[1] In questions of environmental degradation, however, scientific input takes the form of explicit scientific advice to a much lesser extent. Moreover, this lack of explicit scientific advice makes it increasingly difficult to identify what to look for in policy decisions in order to determine questions of effectiveness. A lack of explicit scientific advice may imply that the policy implications of scientific knowledge are not clearly and unambiguously spelled out. When, exactly, scientific knowledge can be considered to have been "acted upon," therefore, is not always a clear-cut judgement. In the decision-making process, policymakers may respond to scientific knowledge at highly varying levels of ambition – from merely taking note of the knowledge brought forward and perhaps preparing a judicial and institutional apparatus to develop agreement on eventual actions at a later stage, to developing policy decisions that, if and when implemented, would represent a solution to the problem. As a minimum requirement then, a decision should, when implemented, represent some sort of behavioural change directed towards the resolution of the problem in question for a decision to qualify as "acting upon" scientific knowledge.

Second, scientific knowledge has a powerful legitimising function for policy arguments (Litfin, 1994). Especially in situations characterised by a high level of scientific uncertainty and a high level of political conflict, policymakers will attempt to couch their policy arguments in scientific terms and also to delegitimise the scientific basis of the arguments of their opponents (see also chapter 3, section 3.4). In such cases, policymakers do not only turn to scientists in search of "truth" or knowledge, but also in search of arguments to legitimise positions into which they already are deeply entrenched (Underdal, in Andresen et al., in press). The fact that a

1 Even in these cases, however, judgements about the effectiveness of the science–policy dialogue on the basis of policymakers' adoption of scientific advice has become problematic particularly because of the conflicts of value increasingly associated with the management of marine resources (especially whaling). (See Andresen in Andresen et al, in press).

policy decision seems to reflect a scientific knowledge base does not, therefore, necessarily imply that the science–policy dialogue has been effective. It could simply imply that the "winning" party or coalition was more successful in the legitimisation–delegitimisation process of policy arguments. The significance of this effect can also be assumed to increase with the trend towards scientific assessments rather than explicit scientific advice, since assessments may leave more room for (individual) interpretation of scientific evidence, thus increasing the potential power of those capable of utilising the legitimisation–delegitimisation mechanism for manipulation. The outcome in such cases, therefore, could be explained by policymakers' opportunistic adoption of scientific findings rather than an effective science–policy dialogue. How to distinguish between these outcomes is, however, an almost insurmountable methodological challenge. We can never truly know an actors' motivations for choosing a particular course of action, although the likelihood that a decision reflects a genuine acceptance of a knowledge base may at least be indicated by analysis of the political context within which the decision is made.

Finally, and perhaps most importantly, an opposite case where the science–policy dialogue has functioned well but where science still to a very little extent is reflected in the decisions made is perfectly conceivable. This may be so because of a host of other factors, not directly linked to the science–policy dialogue, that affect the extent to which policymakers actually respond to scientific findings in terms of behaviour change. The effectiveness of the science–policy dialogue can contribute to "rational" policymaking only up to a certain point. At the end of the day, the political setting (distribution of power, level of conflict, etc.), media coverage, the actions and political power of environmental NGOs, and a host of other factors, will have significant influence over whether or not policymakers choose to respond to available scientific knowledge in terms of action. Thus, policymakers' inaction may well be explained by factors other than the malfunctioning of the science–policy dialogue. The inability of this conceptualisation of effectiveness to capture this type of outcome indicates that the extent to which scientific knowledge has been acted upon by policymakers is insufficient as a sole indicator of the effectiveness of the science–policy dialogue.

Hence, while the acted-upon conceptualisation of effectiveness can be argued to capture the essential (official) objective of science–policy interaction – a reflection of scientific knowledge in political decision-making – it does not provide us with satisfactory analytical tools to make judgements about different levels of effectiveness. In particular, this conceptualisation suffers from the inability to capture processes where the science–policy dialogue has functioned well, but where science still, for reasons beyond which the effectiveness of the science–policy dialogue can make a difference, is not well reflected in the decisions made.

2.3 Policy Acceptance as Reflected in a Consensual Problem Diagnosis

The inadequacy of the acted-upon conceptualisation of effectiveness to handle the problems discussed above may be addressed by including an additional indicator focusing on the extent to which the knowledge base provided by scientists is accepted by policymakers as factually valid and scientifically authoritative. Policymakers' acknowledgement of the scientific authority of the knowledge base and the factual validity of its substantive conclusions is a logical precondition for actions. Thus, the effectiveness of a science–policy dialogue is not only a function of the extent to which scientific advice is acted upon, but also the extent to which policymakers have accepted the knowledge base as a factually valid and legitimate guide to action. Acceptance of (the validity of) scientific findings may thus be employed as a supplementary, lower-level indicator of effectiveness in science–policy interaction.

In terms of operationalisation, policymakers' acceptance of a knowledge base is assumed to be reflected in the extent to which a problem diagnosis, consensual among both scientists and policymakers, has been developed. A consensual problem diagnosis is seen as one product of an effective science–policy dialogue. Thus, an analytical distinction between the outcome of the policymaking process as a whole (the reflection of scientific findings in policy decisions) and the outcome of the sub-process of science–policy interaction (policymakers' acceptance as reflected in the provision or non-provision of a consensual problem diagnosis) is introduced. This conceptualisation of effectiveness has several advantages.

First, in cases where scientific knowledge is not reflected in policy decisions, this conceptualisation allows us to distinguish between explanations related to the (in)effectiveness of the science–policy dialogue and those related to other factors. This conceptualisation also allows us to explore the limits to the potential impact of the science–policy process on policy outcomes. And finally, since this conceptualisation more fully takes into account the complexity and interactive nature of the science–policy process, it also allows us to identify the pitfalls that may impede the effectiveness of the process.

2.3.1 The Development of a Consensual Problem Diagnosis

Especially with the introduction of the new environmental problems on the international political agenda, scholarly attention has been redirected from the processes whereby scientific knowledge is communicated to policymakers towards the processes scientific knowledge is integrated with policy concerns in comprehensive problem definitions that form interpretative frameworks whereby problems are made negotiable (see especially Hajer, 1995). Several studies indicate that policymakers' general capacity to take knowledge into account, and more specifically their perceptions and interpretations of this knowledge, is highly dependent upon how the piece of information is or can be related to their interests in the particular area in question (see for instance, Litfin, 1994; Jasanoff, 1990; Hajer, 1995). Information that does not "fit" into policymakers' interpretative frameworks – their political concerns and perceptions of reality – may be disregarded. "Old" knowledge may thus reappear as "new" knowledge once an interpretative framework whereby the piece of information can be linked to interests and goals is provided. (For an example of this, see Litfin's (1994) discussion of the chlorine loading concept in the case of ozone depletion.) Moreover, this mechanism implies that in cases where policymakers have different perceptions of reality, the same piece of information may be ascribed different meanings. Similarly, the link to policymaking has also been shown to affect the way scientists formulate and frame pieces of scientific information and bodies of knowledge (see especially Shackley and Wynne, 1997). This indicates that a scientific knowledge base provided in the context of policymaking both shapes, and is itself shaped by, policy (Jasanoff, 1990). It should be emphasised that this "feedback" effect, which suggests that science itself is shaped by policy, is not considered to take the form of an explicit and detectable bias of scientific information in favour of particular interests, but should rather be regarded as a result of the interactive dynamics taken on by the diagnostic process. An approach focusing on the development of problem diagnoses thus suggests that in interaction with policy, science departs from the domain of "pure research" and takes one step into the political domain. It also suggests that a policymaking process should not simply be regarded as a problem-solving process, it also contains an element directed towards the construction of problems in ways that make problems "solvable" (Hajer, 1995). This approach thus envisages a zone within which science and policy interacts and where the linkage between facts and values is defined and redefined in the development of consensual problem diagnoses constituting interpretative frameworks for policymaking and problem solving. A consensual problem diagnosis, then, may be understood as the linking of facts (scientific knowledge) and values (policy goals) into an interpretative framework whereby a problem is defined in terms of its basic cause-and-effect relationships *and* alternative paths to its solution are suggested.

More specifically, a consensual problem diagnosis is assumed to be constituted by two main elements: (i) A knowledge base that describes the basic cause-and-effect relationships of the problem area in question, and whose representativenss of state-of-the-art knowledge within relevant fields of research is acknowledge by scientists as well as policymakers; and (ii) A consensus among both scientists and policymakers on how this body of knowledge is linked to policy goals.

The first constitutive element does not necessarily imply that state-of-the-art knowledge represented in the knowledge base itself is consensual within the scientific community: it is its *representation* within the policymaking context that is consensual among all "stakeholders" (scientists as well as policymakers). Hence, policymakers to some extent take part in discussions regarding how specific pieces of information should be interpreted and framed within the context of specific problems. Discussions on appropriate interpretations and framing of knowledge may take place in association with discussions on how a piece of information is linked to policy goals. Hence, we assume that the two constitutive elements of a consensual problem diagnosis may, but need not, be developed simultaneously. In contrast to an "ordinary" scientific process, policymakers have a say with regard to appropriate interpretations of a body of scientific knowledge, and scientists have a say with regard to how a body of scientific knowledge is linked to policy goals. In the course of an effective science–policy dialogue, therefore, scientific knowledge and policy goals are linked into a comprehensive problem diagnosis that constitutes the basis for further political negotiations in the development of political solutions.

The second constitutive element of a consensual problem diagnosis does not imply an assumption of consensus regarding the prioritisation of policy goals, nor does it imply an assumption of consensus regarding what constitutes the "best" solution to a common problem. It does, however, imply that major political concerns are addressed – that the policy implications of a body of knowledge are at least tentatively suggested and that a range of alternative solutions is indicated, although this may not have been made explicit.

Scientists are not assumed to play an active role in the development and design of political solutions, but difficulties in coming to terms on political solutions may imply a redefinition of the problem. In some cases, however, science-based models that facilitate decision-making by serving as decision-making tools may constitute one element of the knowledge base provided. In such cases, scientists' participation in negotiations over details may be somewhat increased. Even in such cases, however, scientists' participation in this phase of the policymaking process is assumed to be limited.

It is not assumed that scientific consensus necessarily leads to political consensus, neither with regard to the way knowledge is linked to policy goals nor with regard to the priority of goals and choice of solutions. Thus, a situation where a consensual problem diagnosis has been developed but where this is not reflected in the decisions (or non-decisions) made is perfectly conceivable. It is assumed, however, that the development of a consensual problem diagnosis may facilitate negotiations on solutions to the extent that it, for instance, directs policymakers' attention to potentially integrative solutions in situations where these otherwise would not have been discovered (integrative bargaining is discussed in chapter 3, section 3.3.2), or to the extent that it provides decision-making tools that enable policymakers to deal with difficult decision-making problems. In this sense, the problem can be regarded as having been "constructed." When the construction of a problem rests on the broadest possible consensus on the problem diagnosis, it is not easily deconstructed by political opponents (non-policymaking actors such as environmental or industrial NGOs, or policymakers in opposition at the national level).

2.4 Towards a Comprehensive Definition of Effectiveness

A conceptualisation of effectiveness has been suggested in terms of three levels: (i) the development of a knowledge base, representative of state-of-the-art knowledge in relevant fields of research, (ii) a (tentative) suggestion of linkages between this knowledge base and valued policy goals, and (iii) the adoption and utilisation of this problem diagnosis as a premise for policy choice.

These three levels do not constitute a cumulative scale. While level one is a logical precondition for levels two and three, level two does not always precede level three. A science–policy dialogue effective in terms of level one and three, but not two, is therefore perfectly conceivable. A science–policy dialogue cannot, however, be effective in terms of either level two or level three without level one. This conceptualisation of effectiveness is illustrated in tables 2.1 and 2.2.

Table 1. The effectiveness of processes of science–policy interaction in the absence of a consensual problem diagnosis

The knowledge base is acted upon	**The factual validity of the knowledge base is accepted:** **No**	**Yes**
No	The science–policy dialogue has been ineffective. (Effectiveness at level 0).	Effectiveness at level 1.
Yes	Scientific knowledge has been acted upon for other reasons than policymakers' acceptance of the factual validity of the knowledge base.	Effectiveness at level 3.

Table 2. The effectiveness of processes of science–policy interaction judged in terms of the extent to which policymakers accept and act upon a consensual problem diagnosis

A problem diagnosis has been developed	**Policymakers have adopted the scientific knowledge base in terms of**	
	Acceptance (of factual validity)	**Actions** (acceptance of "policy implications")
No	Effectiveness at level 1. (A consensual knowledge base)	Policymakers have acted upon scientific knowledge in the absence of a consensual problem diagnosis. The situation may be analysed according to table 2.1.
Yes	Effectiveness at level 2. (A consensual problem diagnosis)	Effectiveness at level 3. (The consensual problem diagnosis has been acted upon by policymakers)

2.5 A "Consensual Problem Diagnosis" versus "Consensual Knowledge"

The notion of consensual problem diagnosis developed here has its origin in the notion of "consensual knowledge", which has become a central focus of attention among political scientists studying the role of science in international policymaking. It is a deliberate choice not to employ this term, however, because of its conceptual ambiguity.

"Consensual knowledge" surfaced as a key concept in the study of international relations particularly in the works of John Gerard Ruggie and Ernst B. Haas et al. around the mid-1970s (J. G. Ruggie, 1975; E. B. Haas, 1975; E. B. Haas et al., 1977). During the latter part of the 1980s and the beginning of the 1990s, the concept of consensual knowledge has gained increased salience among scholars of international negotiation and regime theory, especially within the field of international environmental management. This development is not least due to the work of Peter M. Haas and the development of the approach that has come to be known as the "epistemic communities approach" to international regimes (1989a,b; 1992a,b; 1993a,b).

The notion of consensual knowledge has gained acceptance within international relations literature to the extent that it has become a household word among scholars working in this field. Despite its common usage, however, the substantive meaning of the term has been given little attention. The concept is often poorly defined, or attributed different properties by different authors. Clear conceptual borders between "consensual knowledge" and neighbouring concepts such as "scientific consensus", "shared knowledge", and "ideas" are not well developed. This is evident in the literature, where these concepts are often used interchangeably. The following paragraph is illustrative in this regard:

> Given their technical nature, it is clear that tackling environmental problems will always require a considerable input of *(scientific) knowledge*. The identification of an ecological problem presupposes *such consensual knowledge. ... Shared knowledge* is therefore a necessary condition of regime formation, ... (List and Rittberger, 1992:103, emphasis added.)

One source of this conceptual ambiguity may be found in Ernst B. Haas's own employment of these concepts. Concepts such as "scientific notions," "scientific knowledge", "knowledge", and "consensual knowledge" are seemingly understood in different terms, but without a sufficiently clarifying guide as to how these terms differ and how they are assumed to be related to each other (E. B. Haas et. al., 1977; E. B. Haas, 1980; 1990). "Knowledge" is defined by Ernst Haas as "the sum of technical information and theories about that information which commands sufficient consensus at a given time among interested actors to serve as a guide to public policy designed to achieve some social goal" (1980: 368, emphasis deleted). Knowledge, therefore, is consensual by definition, and the conceptual borders between "knowledge" and "consensual knowledge" thus become difficult to grasp. Similarly, Ernst Haas's employment of the term "consensual knowledge" seems to build on an assumption of "scientific consensus" (understood, for instance, as disciplinary consensus, i.e., a consensus within scientific disciplines). In his depiction of four possible world-order models resulting from learning processes, it becomes clear that an assumption of disciplinary consensus is the basis for all four (E. B. Haas et. al., 1977). He maintains, for instance, that "while the experts may disagree on how to link [a set of activities], they are in considerable accord on how to deal with each one singly. In other words, 'not more consensus' does not mean 'no consensus' " (Haas et. al., 1977: 54). Thus, as argued by Litfin, "the authors' discussion is really about the extent to which interdisciplinary linkage is present. In fact, all four world-order models, even the most disorderly of them, presume that expert consensus exists at least on single issues" (Litfin, 1994: 44). Hence, Ernst Haas's conception of "consensual knowledge" seems to imply something more than mere scientific consensus, namely an explicit linkage to, and also linkage of, policy goals. This distinction is, however, not made explicit. With the synonymous employment of "knowledge" and "consensual knowledge", the conceptual ambiguity becomes significant.

This conceptual ambiguity is reinforced in Peter M. Haas's work on "epistemic communities" (1989a,b, 1992a,b, 1993a,b). In Peter Haas's work, "consensual knowledge" constitutes the glue of "epistemic communities." An epistemic community is defined as "a network of professionals with recognized expertise and competence in a particular domain and an authoritative claim to policy-relevant knowledge within that domain or issue-area" (1992a: 3). Epistemic communities are distinguished from other groups in terms of their "shared set of causal and principled (analytic and normative) beliefs, a consensual knowledge base, and a common policy enterprise (common interests)" (1992a: 18). The ambiguities in the distinction between "shared causal and principled beliefs" and "consensual knowledge" notwithstanding, again we get the clear impression that consensual knowledge is something more than scientific consensus. However, "consensual knowledge" is also used interchangeably with "scientific convergence" (1993a:176; Young and Osherenko, 1993: 19). Peter Haas assumes that the consensual nature of the knowledge shared by experts constituting an epistemic community is the main mechanism whereby experts influence policymaking. This assumption is not subject to testing. Thus, Peter Haas discusses how the consensual knowledge base is "diffused" to policymakers by members of the epistemic community's "capturing" of important bureaucratic positions (1993b: 179) and "political infiltration" into governing institutions (1993b: 179), from which they may "usurp" control over decision-making (1992a: 27). While Ernst Haas defines consensual knowledge as knowledge "accepted as a basis for public policy by groups and individuals professing varying political ideologies" (1980: 368), Peter Haas erects a distinction between the epistemic community, united by a consensual knowledge base, and policymakers. However, since an epistemic community also may include policymakers, the distinction between an epistemic community and policymakers itself becomes highly ambiguous.

Provided that the authors consider a consensual knowledge base to be constituted by a linkage between scientific knowledge and policy goals, the process whereby this consensus is developed is not made subject to analysis. This indifference to the way consensual knowledge is developed leaves us with the impression that scientific consensus is assumed to lead to political consensus (consensus on policy goals) and that a "consensual knowledge base," whereby the two are linked, simply "emerges."

The concept of a consensual problem diagnosis developed here differs from the notion of consensual knowledge in at least two important respects: First, and most importantly, a consensual problem diagnosis may be, but often is not, based on a scientific consensus: It is the *representation* of state-of-the-art scientific knowledge that is consensual – with the identification of its inherent scientific uncertainties and controversies – not necessarily the scientific knowledge itself. Second, in this conceptualisation of effectiveness, the development of a consensual problem diagnosis is seen as the outcome of a process of science–policy interaction (effective at level 2). Hence, we do not assume that this consensus "emerges" by itself. The nature of the process whereby this outcome is brought about and the factors that affect its provision is the subject of study in this analysis.

2.6 In Sum

Effectiveness in processes of science–policy interaction is understood in terms of the extent to which policymakers accept and subsequently act upon the knowledge base provided by the scientific community. Policymakers' acknowledgement of the scientific authority of the knowledge base and their acceptance of its substantive conclusions as factually valid is assumed to be reflected in the extent to which a problem diagnosis – consensual among both scientists and policymakers – is developed. A consensual problem diagnosis is seen as being constituted by two main elements: (i) A knowledge base describing the basic cause-and-effect relationships of the problem area in question, whose representativeness of state-of-the-art knowledge within relevant fields of research is acknowledged by both scientists and policymakers; and (ii) A consensus among both scientists and policymakers on how this body of knowledge is linked to policy goals. The development of these two elements constitutes the first two levels in the three-level conceptualisation of effectiveness. At the third level, policymakers accept also the policy implications of the knowledge base and adopt this problem diagnosis as a premise for the decisions made. It is emphasised that these levels do not constitute a cumulative scale. While level one is a (logical) precondition for levels two and three, level two may not necessarily precede level three. Policymakers may therefore act upon scientific knowledge in the absence of a consensual problem diagnosis as long as they have accepted the factual validity of scientific findings in accordance with level one.

Chapter 3

The Science–Policy Nexus

3.1 Introduction

Science and politics constitute two distinct systems of behaviour, each with its own purpose, constitutive norms and internal logic. Science is (ideally) conceived of as a truth-seeking endeavour whose norms and guidelines for behaviour are directed towards the generation of impartial and disinterested knowledge. Politics constitutes a system for the generation of (collective) decisions, where behaviour is directed towards the realisation of (individual) rational interests in these decisions. In contrast to the ideal of impartiality characterising the scientific method, political behaviour is characterised by a strategic reasoning where the instrumental utilisation – as well as manipulation and distortion – of knowledge may constitute central elements in political strategies whereby individual interests are sought realised.

The relationship between science and politics, therefore, is characterised by an immanent tension. Some may even characterise their relationship as one of opposite poles where science is everything politics is not: pure, objective, subject to rational analytical reasoning and thus not hostage to manipulation tactics and coercive power – ingredients often associated with politics. While analyses show that the relationship between science and politics is far from being as clear-cut as this image suggests, this image has a strong position in the public mind as well as among practising scientists and policymakers themselves. Any dialogue between these two systems of behaviour takes place in the shadow of this image, and the immanent tension it creates may represent an obstacle to its effectiveness.

This analysis explores some of the inherent pitfalls of the science–policy dialogue associated with international environmental policymaking and investigates some of the means available to enhance its effectiveness. In particular, the analysis investigates the extent to which and how the design of the institutional framework within which the science–policy dialogue takes place may represent a potential instrument in this regard. While the impact of institutions on science and politics – as separate endeavours – has been explored both empirically and theoretically within sociological studies of science and political science, respectively, there is no theoretical apparatus readily available for analysing the impact of this factor on processes in which these systems of behaviour interact. In this chapter, therefore, a theoretical approach that draws upon a broad spectrum of fields and disciplines – science philosophy, sociology of science, decision theory and negotiation theory – is developed to analyse the nature and dynamics of processes of science–policy interaction (section 3.4 in this chapter) and to determine the extent to which institutional arrangements may be utilised as facilitating devices to enhance the effectiveness of the interaction (analysed in chapter 4).

While bearing elements of both science and politics, the nature and dynamics of science–policy interaction processes emerge from the interaction itself when these two distinct systems of behaviour meet. The point of departure for the development of a theoretical approach which captures the nature and dynamics characterising this interaction is the internal dynamics of science and politics in their "pure" forms. Thus, section 3.2 is devoted to an investigation of the internal dynamics of the process whereby science produces knowledge. In section 3.3 we do a similar analysis of the process whereby politics produces decisions. Finally, the nature of the dynamics that are generated when these two systems of behaviour interact is discussed in section 3.4.

3.2 The Internal Dynamics of Scientific Inquiry

Identifying the nature of the internal dynamics that characterise the process whereby scientific knowledge is produced is a complex task that leads us straight into the more fundamental question of what "science" is. Each of the meta-scientific disciplines (philosophy, sociology, psychology, anthropology, etc., all with the postscript "of science") have formulated their own answers emphasising different features and concentrating on different aspects of science – each with a tendency to treat its own definition of science as self-sufficient (see Ziman, 1984). "In truth", Ziman argues,

> science is all of these things, and more. It is indeed the product of research; it does employ characteristic methods; it is a body of organized

knowledge; it is a means of solving problems. It is also a social institution; it needs material facilities; it is an educational theme; it is a cultural resource; it requires to be managed; it is a major factor in human affairs. Our "model" of science must relate and reconcile these diverse and sometimes contradictory aspects. (1984:2)

The traditional view of science portrays research as a rational, rule-governed process, in which the implementation of the scientific method is the main mechanism by which established knowledge is distinguished from mere knowledge-claims. According to this view, the scientific method constitutes the common framework within which hypotheses and theories are developed and tested. The scientific method also functions as the main arbiter of disputes over what is "true". "Knowledge" is what the community of scientists holds to be true on the basis of extensive scrutiny in accordance with demanding and discriminating methods. One important implication of this line of reasoning is that, in order to qualify as knowledge, a proposition must be consensual or inter-subjective in the sense that any competent scientist, applying the scientific method correctly, would reach the same conclusion. Should scientific dispute occur, there are only two possible explanations: either one of the dissenters is (value) biased (i.e., not genuinely seeking the truth), or the scientific method is applied erroneously (see, for instance, Collingridge and Reeve, 1986). Accordingly, ("true") knowledge can – with some reservations – be distinguished from mere knowledge claims by the operational criterion of consensus within the scientific community.

The traditional view of science – particularly the proposition that knowledge depicted in consensual terms carries more political weight than knowledge not depicted in these terms – has a strong position among political scientists who investigate the role of science in policymaking (see, for instance, E. B. Haas et al., 1977; E. B. Haas, 1980; 1990; P. M. Haas, 1989a,b; 1992a,b; 1993a,b; see also chapter 2). It has also been maintained, however, that consensus in science is a political invention alien to science itself and that the scientific community "has no recognised process of consensus creation and would be shocked by the idea" (Boehmer-Christiansen, 1995: 2). Thus the organising questions of this section are first, the role ascribed to consensus in science by science philosophers and sociologists, and second, the nature of the process whereby consensus in science is developed.

3.2.1 The Role of Consensus in Science

Although the traditional view of science has been seriously challenged by philosophers and sociologists of science, scientific consensus is still ascribed a major, although debated, role in theory development and knowledge production. The dispute between three central philosophers of science, Karl Popper, Thomas S. Kuhn and Imre Lakatos, which has taken place since the publication of Kuhn's seminal book *The Structure of Scientific Revolutions* (1962), is representative of the philosophical aspects of this debate.

The major proposition in Kuhn's work is that "normal science" takes place within the framework of "paradigms", i.e., "universally recognized scientific achievements that for a time provide model problems and solutions to a community of practitioners" (Kuhn, 1962/70:viii[1]). Paradigms, according to Kuhn, do not merely inform the work of scientists, they constitute the "particular constellation to which the group, at any given time, is committed" (ibid.: 4) and, constituting the foundation for research, they are a prerequisite for effective research;

> Effective research scarcely begins before a scientific community thinks it has acquired firm answers to questions like the following: What are the fundamental entities of which the universe is composed? How do these interact with each other and with the senses? What questions may legitimately be asked about such entities and what techniques employed in seeking solutions? (ibid:5)

The answer to questions such as these are "firmly embedded in the educational initiation that prepares and licenses the student for professional practice. Because that education is both rigorous and rigid, these answers come to exert a deep hold on the scientific mind" (ibid.). According to Kuhn, these answers "exert a hold on the scientific mind" to the extent that research conducted within the framework of paradigms can be described as "a strenuous and devoted attempt to force nature into the conceptual boxes supplied by professional education" (ibid.). It is only occasionally, in the event of anomalies' "stubborn refusal to be assimilated to existing paradigms" (ibid.: 97), that new theories and new paradigms are developed, and these situations constitute "scientific revolutions".

Karl Popper disputes the picture of science painted by Kuhn. While Kuhn regards scientific consensus, or "normal science", as a necessary condition for effective research, Popper regards the phenomenon described by Kuhn as a "danger to science": "The 'normal' scientist, as described by Kuhn, has been badly taught. He has been taught in a dogmatic spirit: he is a victim of indoctrination" (Popper, 1970:52–53). Taking a closer look at Popper's arguments, however, we find that it is not consensus as such he disputes, but

[1] This is the initial definition provided by Kuhn. It should be noted that the concept is ascribed several meanings throughout Kuhn's text. Masterman, for instance, has traced no less than 21 definitions of the term in Kuhn's own writing (Masterman, 1970:61–65). For a response from Kuhn to these criticisms, see Kuhn, 1977a.

an explicit aim on the part of the scientist to verify rather than falsify existing theories:

> ...the dogmatic attitude is clearly related to the tendency to *verify* our laws and schemata by seeking to apply them and clarify them, even to the point of neglecting refutations, whereas the critical attitude is one of readiness to change them – to test them; to refute them: to *falsify* them, if possible. This suggests that we may identify the critical attitude with the scientific attitude, and the dogmatic attitude with the one which we have described as pseudo-scientific. (Popper, 1963:50)

Popper, too, recognises that scientists operate within the framework of some sort of consensus, or "structures of scientific doctrines":

> A scientist engaged in a piece of research, say in physics, can attack his problem straight away. He can go to the heart of the matter: that is, to the heart of an *organized structure. For a structure of scientific doctrines is already in existence; and with it a generally accepted problem-situation.* This is why he may leave it to others to fit his contribution into the framework of scientific knowledge. (1968:13, emphasis in original).

An important point to Popper, and in contrast to Kuhn, is that the individual scientist can always break out of this organised structure of knowledge:

> I do admit that at any moment we are prisoners caught in the framework of our own theories; our expectations; our past experience; our language. But we are prisoners in a Pickwickian sense: if we try, we can break out of our framework at any time. Admittedly, we shall find ourselves again in a framework, but it will be a better and roomier one; and we can at any moment break out of it again. (1970:56)

Thus, we find that Popper and Kuhn are in some sort of agreement on the role of "consensus" in science (though neither of them employ this term), while they are in considerable disagreement with regard to the appropriate attitude of the individual scientist towards this framework, and his or her incentives and capacity to break out.[2]

Lakatos sees scientific growth in terms of series of theories. It is not one single theory, but a series of theories that is appraised by the scientific

[2] This has given rise to an interpretation of the disagreement between the two as rooted in methodology: It has been claimed that Popper represents a *normative* approach, whereas Kuhn represents a *descriptive* approach. That is, Kuhn says something about how things are, while Popper says something about how things ought to be (see, for instance, Watkins, 1970). Kuhn, however, does not applaud this interpretation of their dispute (see 1970b: 233).

community as "scientific" or "pseudo-scientific" (Lakatos, 1970). These series of theories are characterised by continuity. The mechanism providing this continuity, according to Lakatos, is research programmes: "This *continuity* – reminiscent of Kuhnian 'normal science' – plays a vital role in the history of science; the main problems of the logic of discovery cannot be satisfactorily discussed except in the framework of a *methodology of research programmes*" (ibid.: 132, emphasis in original). A research programme in Lakatos's conceptualisation consists of methodological rules that, if followed, will further rational theory development. Some of these rules "tell us what paths of research to avoid (*negative heuristic*), and others what paths to pursue (*positive heuristic*)" (ibid.). All scientific research programmes may be characterised in terms of a "hard core" – which constitutes the theoretical foundation for research within the programme; the basic theories and assumptions made – and a "protective belt". The negative heuristic of the research programme forbids scientists to question or test the basic assumptions that comprise the hard core. "Instead," Lakatos argues, "we must use our ingenuity to articulate or even invent 'auxiliary hypotheses', which form a *protective belt* around this core, and we must redirect the *modus tollens* to *these*" (ibid.: 133). Thus, Lakatos conceives of scientific knowledge in terms of a structure of theories, or a body of knowledge, that is generally accepted and not subject to testing: "This 'core' is 'irrefutable' by the methodological decision of its protagonists: anomalies must lead to changes only in the 'protective' belt of auxiliary, 'observational' hypotheses and initial conditions" (ibid.). The aim of science, then, is to strengthen the hard core by testing, adjusting or replacing the auxiliary hypotheses consituting the protective belt: "A research programme is successful if all this leads to a progressive problemshift; unsuccessful if it leads to a degenerating problemshift" (ibid.).

While the negative heuristic of a research programme specifies its basic, "irrefutable" assumptions, the positive heuristic "consists of a partially articulated set of suggestions or hints on how to change, develop the 'refutable variants' of the research-programme, how to modify, sophisticate, the 'refutable' protective belt" (ibid.: 135). The positive heuristic of a research programme helps scientists to deal with anomalies. The main point emphasised by Lakatos in this regard is that scientists deal with theoretical model simulations of nature, and that "anomalies" or observational refutations of the assumptions upon which such models build, may be irrelevant and are often foreseen by scientists when developing the model:

> A "model" is a set of initial conditions (possibly together with some of the observational theories) which one knows is bound to be replaced during the further development of the programme, and one even knows, more or less, how. This shows once more how irrelevant "refutations" of

> any specific variant are in a research programme: their existence is fully expected, the positive heuristic is there as a strategy both for predicting (producing) and digesting them. Indeed, if the positive heuristic is clearly spelt out, the difficulties of the programme are mathematical rather than empirical. (ibid.: 136, emphasis deleted)

Hence, according to Lakatos, refutations in the protective belt are of a theoretical nature; assumptions and models are not refuted as a consequence of observational anomalies, or 'data', only. In this regard, however, it is not entirely obvious what distinguishes Kuhn and Lakatos, a point also commented upon by Kuhn (see 1970b: 238–41).

The notion of consensus in science has a strong position also among sociologists of science. Mulkay, for instance, maintains that

> It is clear that there is a connection between the idea of "consensus" and that of "knowledge". Only propositions which command (more or less?) universal assent on the part of competent judges can be said to be true "knowledge"; ... As soon as intellectual disagreement appears we seem to be dealing not with demonstrated knowledge, but merely with knowledge-claims. (1978:109)

Similarly, Gilbert suggests that "research findings only become accepted as constituting knowledge following their approval by a research network as a whole" (1976:282). In a similar language, the consensus aspect of scientific knowledge is also emphasised by Ziman when he maintains that the facts and theories of scientific knowledge

> must survive a period of critical study and testing by other competent and disinterested individuals, and must have been found so persuasive that they are almost universally accepted. The objective of Science is not just to acquire information nor to utter all non-contradictory notions; its goal is *consensus* of rational opinion over the widest possible field. (1968: 9)

The notion of consensus in science as an important feature of "established" or "core" knowledge thus seems to be firmly embedded within both the philosophy and the sociology of science. Our discussion suggests that scientific inquiry is understood as being based on a framework of scientific theories and propositions – comprising the fundamental assumptions in the research field – that are generally accepted by the scientific community, and that this body of knowledge is distinguishable from the body of theories and propositions that are not endorsed and appraised by the community. Thus, a common understanding seems to have prevailed in which it is recognised that scientists do operate, at any given time, within frameworks of generally accepted theories and knowledge propositions, comprising some sort of "consensus". It is also recognised, however, that these frameworks of corroborated theories and knowledge propositions are continuously changing, either as the result of an evolutionary process, or in abrupt "revolutions", implying that what constitutes generally accepted scientific knowledge at point *T* does not necessarily do so at point T_1.

There is significant disagreement, moreover, regarding the mechanisms by which consensus is developed – that is, the nature of the process by which knowledge claims are transformed into established knowledge.

3.2.2 The Development of Consensus in Science

Much of the debate between Popper and Kuhn concerns the dynamics characterising theory development in science, or the development of paradigms/organised structures of scientific doctrines, here conceptualised as "scientific consensus". Kuhn has been criticised for introducing what has been referred to as an element of "irrationality" into the research process, when arguing that the criteria for choosing one theory over another are not necessarily those we usually associate with science (for instance, scientific evidence). According to Kuhn the transformation from one paradigm to another is not, as for instance argued by Popper, entirely based on logical scientific reasoning and scientific evidence for the one theory's supremacy over another. In this choice, Kuhn argues, there is a significant element of persuasion. Such a transformation is an extraordinary situation, in which extraordinary means are employed. In normal science, the paradigm guides scientists' choice of problems, methods and solutions, and thus ensures rationality in the research process. In this extraordinary situation, however, it is the paradigm itself that is on trial. Hence, there are no guiding principles and rules for scientists to adhere to. The foundation for rational research itself is lacking, and, accordingly, there are no universal criteria for making the choice. In this situation, therefore, non-scientific factors also determine scientists' choice:

> ...when asked about persuasion rather than proof, the question of the nature of scientific argument has no single or uniform answer. Individual scientists embrace a new paradigm for all sorts of reasons and usually for several at once. Some of these reasons ... lie outside the apparent sphere of science entirely. Others must depend upon idiosyncrasies of autobiography and personality. (1962/70:152–3)

To Kuhn's thesis, Popper has commented that, "the Myth of the Framework is, in our time, the bulwark of irrationalism" (1970:56). Similarly, Lakatos has maintained that, "in Kuhn's view scientific revolution is irrational, a matter for mob psychology" (1970:178), and that, "if ... there is no other way of judging a theory but by assessing the number, faith and vocal energy of its supporters, then ... truth lies in power" (ibid.:93). Kuhn, on the other hand, claims that rather than viewing science as "irrational", he regards it as "arational"[3]: In certain occasional situations, there is no foundation for making a rational choice in the strict sense; in the same situation different choices can be made without any of them demonstrably breaking a common norm, and without any of them being demonstrably "irrational" (in the sense of being occasional, seemingly unmotivated). Kuhn has, for instance, pointed out that what he has emphasised as important values in science correspond to the criteria traditionally emphasised as important in situations of theory choice (1977b). He then points to five criteria by which the adequacy of a theory should be evaluated (the list is not meant to be exhaustive): Accuracy, consistency, scope, simplicity and fruitfulness. He further points out that "I agree entirely with the traditional view that [these criteria] play a vital role when scientists must choose between an established theory and an upstart competitor. Together with others of much the same sort, they provide *the* shared basis for theory choice" (ibid.: 322). He then goes on to argue, however, that there are two problems associated with these criteria (and others of the same sort). That is, first, that they are imprecise, allowing for varying interpretations between individual scientists, and second, that they may be in conflict with one another, and thus not provide clear guidance to the individual scientist. There are no universal rules telling scientists how to choose, or prioritise between, for instance, scope and simplicity in situations when these norms are in conflict. These two features – imprecision and internal contradictions – give leeway for sociological, psychological and idiosyncratic differences in the individual scientist's emphasis on the one criterion over another in a given situation (Kuhn, 1977b). Moreover, Kuhn has argued that accusations of his vagueness with regard to criteria for theory choice, are equally applicable to Popper' thesis:

> What is vague ... about my position is the actual criteria (...) to be applied when deciding whether a particular failure in puzzle-solving is or is not to be attributed to fundamental theory and thus to become an occasion for deep concern. That decision is, however, identical in kind with the decision whether or not the result of a particular test actually falsifies a particular theory, and on that subject Sir Karl is necessarily as vague as I. (1970b: 248)

[3] The source of this distinction is an interview with Kuhn in *Scientific American*: "he still seems pained by the breadth of misunderstanding, by persistent claims, for example, that he thinks scientists are 'irrational'. 'If they had said 'arational', I wouldn't have minded at all', he remarks with no trace of a smile" (*Scientific American*, May 1991:14).

Thus, while Kuhn's thesis implies that non-scientific factors may influence individual scientists' choice of theories in extraordinary situations, non-scientific factors are not ascribed any direct influence on the content of knowledge. That is, within the framework of paradigms, science is regarded as a rational, rule-governed process. Also, when competing paradigms have developed (after "someone" –primarily individuals – have made the choice) the extraordinary in the situation is gradually reduced, and with it, also the leeway for non-scientific influence. The individual choice in extraordinary situations is "only belatedly the unanimous decision of the group" (1970b: 262). When individual paradigm shifts lead to collective paradigm shifts, the theory has been tested and explored to the extent that the risk associated with the transition is negligible. Thus, contrary to many interpretations of Kuhn, it could be argued that while his theory permits non-scientific factors (psychological, sociological) to influence individual scientists' theory choice in occasional, extraordinary situations, it does not suggest non-scientific factors to influence the content of knowledge in "normal science", nor ascribe a major role to non-scientific factors in the process whereby the scientific community evaluates new scientific contributions, any more than Popper's theory does.

The publication of *The Structure of Scientific Revolutions* became a milestone in the history of sociological studies of science in the sense that it opened a whole new field for sociological analysis and thus became a catalyst in the development of the field that has come to be known as the social studies of scientific knowledge.[4] According to this school, Kuhn's analysis convincingly demonstrated that the criteria by which "good" and "bad" science are distinguished are constituted by scientific knowledge itself (see for instance Shapin, 1993), and hence, that the process whereby consensus in science is established and maintained becomes open to sociological analysis (Mulkay, 1978). Building on Kuhn, some sociologists maintain that there is not one, but several social orders in science. Hence, there is not one, but several sets of rules and norms employed by scientists both in the conduct and the evaluation of scientific knowledge (Shapin, 1993). This also has implications for the development of consensus. Thus, Mulkay argues that consensus only occurs under certain conditions: when "scientists share a common scientific/technical background" and "are able to select for attention problems which they judge to be solvable within their common framework" (1978: 111). According to Mulkay, therefore, "scientific consensus, ... is seldom complete and its establishment may well depend on such socio-cultural factors as the selection of problems for which

[4] Some of the propositions associated with this school are not supported by Kuhn himself (see especially Kuhn, 1991).

technically defined solutions appear likely as well as on the application of a relatively uniform scientific perspective" (1978:111).

Another central proposition that has derived from this school of thought is a conception of the internal processes of knowledge production (consensus formation) as processes of *informal negotiations* among scientists (see also Kuhn, 1991). Mulkay, for instance, argues that "scientific consensus in research networks seems to be achieved, at least partly, by processes of informal negotiation between participants who have certain shared as well as certain conflicting interests" (1978:111).

This perspective has also been developed in numerous studies of what goes on in the laboratory – an approach of which Latour and Woolgar's analysis from 1979 is representative. A major point in these studies is to show that social processes with impact on the construction of scientific knowledge are at work not only in "borderline" cases where science interacts with policy in more direct ways, such as for instance when scientific communities are called upon to give advice in processes of policymaking,[5] but also in the routine work of "normal" science in the laboratory. Latour and Woolgar's analysis takes the form of an "anthropological" study of laboratory life. The authors spent two years observing scientists' activities in the laboratory, taking detailed notes of discussions among the scientists and conducted on this basis a detailed analysis of the nature of the scientific process.

Negotiation is, according to Latour and Woolgar, one of several micro-processes whereby facts are *socially constructed*. Their argument, however, is not only that facts are socially constructed, they also maintain that "the process of construction involves the use of certain devices whereby all traces of production are made extremely difficult to detect" (1979/1986:176, emphasis deleted). This approach represents a perspective in which the authors regard the work in the laboratory as representing the "real" science, while the writing up of results obtained in the laboratory for publication represents a *distortion* of the "real" science (see also Cole, 1992). That is, while activities in the laboratory, according to Latour and Woolgar, very much concern the *construction* of facts, scientists' representation in the published accounts of the manner in which the facts were "discovered" serve to conceal the process of construction. Now, this process of concealing the "true nature" of fact construction is, according to Latour and Woolgar, so cleverly done that anyone can be fooled by it:

[5] A situation referred to by Latour and Woolgar as a situation in which, "...one can clearly identify the presence of some politician breathing down the necks of working scientists" (1979:23).

> Once splitting and inversion have occurred [the techniques adopted in order for statements gradually to acquire a "fact-like status"], even the most cynical observers and committed relativists will have difficulty in resisting the impression that the "real" TRF has been found, and that the statement mirrors reality. The further temptation for the observer, once faced with one set of statements and one reality to which these statements correspond, is to marvel at the perfect match between the scientist's statement and the external reality. Since wonder is the mother of philosophy, it is even possible that the observer will begin to *invent* all kinds of fantastic systems to account for this miraculous *adequatio rei et intellectus*. To counter this possibility, we offer our observations of the way this kind of *illusion* is constructed within the laboratory. It is small wonder that the statements appear to match external entities so exactly: they are the same thing. (ibid.:177, emphases added)

The transformation of statements into facts cannot, according to Latour and Woolgar, be explained by "reality" since "it is only after it has become a fact that the effect of reality is obtained. ... It is *because* the controversy settles, that a statement splits into an entity and a statement about an entity; such a split never precedes the resolution of controversy" (ibid.:180, emphasis in original).

In contrast to the approach developed by Latour and Woolgar, Stephen Cole argues that while these studies of "laboratory life" demonstrate certain "irrational" features associated with what Kuhn has labelled the "context of discovery", they fail to demonstrate the impact of these features on the *content* of knowledge, which, in line with that expressed by the authors cited above, refers to communally held knowledge (1992: 102). Hence, by failing to give an account of what happens to the knowledge-claims of laboratory work in terms of *communal acceptance*, Cole argues that these studies also fail to demonstrate any causal link between a social variable and specific cognitive ideas.

Cole sees scientific knowledge in terms of two types of knowledge: knowledge in the "core" and knowledge at the "frontier". The core constitutes the starting point, or the knowledge scientists take as given, providing the basis from which new knowledge is produced. The core is characterised by "substantial consensus", and comprises what we normally refer to as "facts": "knowledge which has been accepted by the scientific community as true or as an adequate representation of nature. In addition, core knowledge must be judged by the community to be 'important' " (Cole, 1992:15). Thus, in order for a new contribution to be permitted into the core, it must be regarded as both true and important. The research frontier, on the other hand, "consists of all the work currently being produced by all active researchers in a given discipline" (ibid.), and in contrast to knowledge in the core, the research frontier is not characterised by consensus. Hence, the crucial variable distinguishing core and frontier knowledge is the presence or absence of consensus.

This distinction has implications for the extent to which consensus is seen as a feature of science: "If we look only at core knowledge and at what scientists say about core knowledge, we will conclude that science is adequately described by the traditional view. If we look at frontier knowledge, however, we will find little confirmation for much of the traditional view" (ibid.: 16). Cole has "measured" the amount of consensus at the frontier by investigating the level of agreement among reviewers in their evaluation of research proposals to the National Science Foundation and finds that "there may not be significantly more consensus in evaluating new scientific ideas than there is in judging nonscientific items such as human beauty, new works of art, or Bordaux wines" (ibid.: 19). He thus finds that the selection of which scientific ideas that are given a research grant and thereby a chance to be further developed largely depends upon sheer luck and is to a very little extent influenced by the content of the proposals. Hence, the *foci of attention* of scientific communities seems to be heavily influenced by subjective or social factors, which further implies that what might have become significant contributions to increased understanding of some phenomenon is left in the dark for non-scientific reasons. It does not, however, imply that the contributions that are accepted into the core are accepted for non-scientific reasons.

As argued by Cole, "all human behavior is conducted in a social context and is therefore influenced by society" (ibid.: 62). The important question, therefore, is not *whether* social variables influence science, but "exactly what types of social variables influence science in precisely what ways" (ibid.: 68). By distinguishing between core and frontier knowledge, it is possible to differentiate the process whereby communal scientific knowledge is developed, and specify in greater detail the scope for social influence at the various junctures of the process. Cole investigates three sets of interactive variables assumed to be at work in the process whereby new scientific contributions enter the core: the content of the work itself, the social characteristics of the authors, and the operation of social processes such as intellectual authority. Of these three sets of variables he finds the first and the last to be most important. With regard to the former, he maintains that "the most important evidence in support of my position is the fact that some discoveries are almost immediately accepted by the scientific community as being true and are quickly added to the core" (ibid.: 23). This rapid acceptance of some work, Cole argues, cannot be fully explained by a constructivist approach.

3.2.3 In Sum

There is broad agreement among philosophers and sociologists that consensus is a central feature of "core" or "established" knowledge. The notion of consensus is well established as the major criterion by which scientists themselves distinguish between established knowledge and knowledge claims. With regard to the mechanisms at work in processes of consensus formation on the other hand, disagreement is profound. Positions on this issue may be conceptualised in terms of a continuum ranging from a view of the process of consensus formation as an entirely norm-and-rule-governed process at the one extreme, to a view of the consensus process as entirely power-and-interest driven at the other. Still, our discussion clearly indicates that the process whereby knowledge is produced is distinctively different from a political process.

Most importantly, our discussion demonstrates that the road to consensus in science is not as straightforward, uncontroversial and transparent as one might assume on the basis of the traditional view in which science is portrayed as a "pure", "objective" and largely "value-free" endeavour. Policymakers may thus be confronted by scientific dissent that may not simply be dismissed as value bias or incompetence on behalf of the scientists. Moreover, our discussion illustrates that even "established" knowledge may be less robust and hence to a larger extent subject to interpretation and more vulnerable to manipulation and distortion than the traditional image of science suggests.

3.3 The Internal Dynamics of Politics

While the purpose of research is to produce knowledge, the purpose of politics is to produce authoritative decisions on behalf of a society or group. In international politics, the most common decision-making procedure is negotiation.

Negotiation is often defined as a decision-making procedure between two or more parties who have some interests that are shared and others that are opposed, and where the one party's utility in the outcome depends upon the other party's course of action (Fisher and Ury, 1981/87). Hence, negotiation is often understood as a means whereby a situation of strategic interaction, or interdependent decision-making, can be handled (Young, 1975). Strategic interaction characterises a situation where the utility of each actor's choice of behaviour is dependent upon the other (relevant) actors' choice of behaviour (Young, 1975:6). Parties may have diverging preferences because their interests are in conflict. Preferences may, however, also differ because the parties have different information and/or opinions about the likelihood that a consequence (desired by all parties) actually will follow from a specific course of action. Parties may thus negotiate in order to co-ordinate not only their behaviour but also their expectations to alternative solutions. For there to be any point in negotiating, there has to be a potential joint gain and at least partly diverging preferences.

Information and knowledge play a crucial role in all negotiations. The parties' handling of and attitude towards information and what kind of information the parties need is, however, at least partly determined by the nature of negotiation dynamics and which phase of the process the negotiations are in. The focus of this section is, therefore, on which implications these factors may have for the role of information and knowledge in the negotiation process. Negotiation dynamics is discussed with a point of departure in Walton and McKersie's seminal book *A Behavioral Theory of Labor Negotiations* (1965/93) and their two models of negotiation dynamics: the distributive and the integrative bargaining models. The manners in which the parties' demand for different kinds of information and knowledge may vary according to negotiation phase is discussed with a point of departure in Zartman and Berman's "formula-detail" approach (1982).

3.3.1 The Strategic Value of Information in Distributive Bargaining

Parties to a negotiation will choose negotiation tactics according to (their perception of) the nature of the issues at stake – whether the amount of the good in question is fixed or variable – and the extent to which their interests are in conflict with the interests of their opponents. By combining these dimensions, Walton and McKersie (1965/93) have developed two models of bargaining. The distributive bargaining model tends to apply in situations where the amount of the good in question is fixed and the parties' interests are in conflict in the sense that what the one gains, the other looses. The integrative bargaining model (which will be discussed in the next section) tends to apply in situations where the opposite is the case – where the amount of the good in question is variable and the parties' interests are identical, purely coincidental or complementary. It should be noted, however, that distributive issues may in some cases be approached by integrative bargaining tactics. Midgaard points out that, "the possibility of identifying new outcomes within a 'distributive' framework (...) can call for a co-operative effort as well as for techniques specific to pure bargaining such as keeping back relevant information" (1976: 122). In between these two extremes we find mixed-motive bargaining, which captures the most common real-world situations where the agenda contains a mixture of conflictual and collaborative items. Hence, negotiation dynamics may be conceptualised in terms of a continuum, with pure distributive bargaining at the one extreme, pure integrative bargaining at the other extreme, and a host of situations in between in which actors are confronted by the dilemmas involved in the choice of complex combinations of distributive and integrative bargaining tactics.[6]

The subject matter of distributive bargaining is *issues*. An issue is characterised by a fixed-sum, variable-share pay-off structure: "It describes a situation in which there is some fixed value available to the parties but in which they may influence shares which go to each" (Walton and McKersie, 1965/93: 13). In its pure form, an issue would require that whatever gains are available to one necessarily entails a corresponding and equal sacrifice by the other. In this regard, the interests of the parties are in conflict.

In negotiations, parties act strategically. In distributive bargaining, each party chooses the course of action (strategy) that maximises his individual gain or minimises his individual loss given the other parties' pay-off structure and hence, strategies. Each party enters negotiations with a hope of achieving a negotiated solution which is better than what they can obtain without a negotiated settlement. Each party thus enters negotiations with some notion of their Best Alternative To a Negotiated Agreement (BATNA) (Fisher and Ury, 1982/87).

A party's BATNA is usually considered to constitute that party's resistance point – the point beyond which he would prefer to terminate the bargaining rather than accept a negotiated solution. Often, however, a party also enters negotiations with a more or less clear idea of what would constitute a satisfactory outcome. And that point may not necessarily match that party's BATNA. Thus, a party may have two benchmarks against which possible solutions are evaluated. In some cases, he may be reluctant to accept a solution alternative which is inferior to his standard of satisfaction even if it is superior to his BATNA (Underdal, 1992: 230). Whether or not he will accept a solution under these conditions will, inter alia, depend upon the costs and benefits associated with making such a move as compared to the utility associated with the direct impact of the move. That is, his decision about whether or not to accept a solution under these conditions depends upon the process-generated stakes with which this acceptance is associated (Underdal, 1992; 1983).

The BATNA (or standard of satisfaction) of each party to negotiations defines the area within which a solution must be found: the settlement range (Figure 3.1). Within the settlement range are the solution alternatives all parties consider to be better than their BATNA (or consider to be satisfactory). The settlement range thus constitutes the set of possible solutions which satisfy the minimum requirements of all parties. If there are

[6] For a further elaboration of these concepts, see also Midgaard, 1976.

no possible solutions within this range, there is no point in continuing the bargaining.

Zeuthen's dynamic bargaining model may serve as a representative illustration of the dynamics that distributive bargaining may take on and the mechanisms by which negotiations converge on one settlement point. Zeuthen conceptualises the bargaining process as a series of successive decision points that confront bargainers with the choice of reiterating their previous demand or making a concession. The criterion rational bargainers use to make this decision is, according to Zeuthen, the risk of conflict. Zeuthen's basic proposition is that parties will not yield or make concessions unless or until the risk of conflict (both parties pursue a hard bargaining strategy) becomes unacceptable. He contends that a party's decision to stick with his demand or to make a concession at any given point in the bargaining process is a function of each party's critical risk. The party with the higher tolerable risk can hold out longer and force the party with the lower tolerable risk to make a concession.

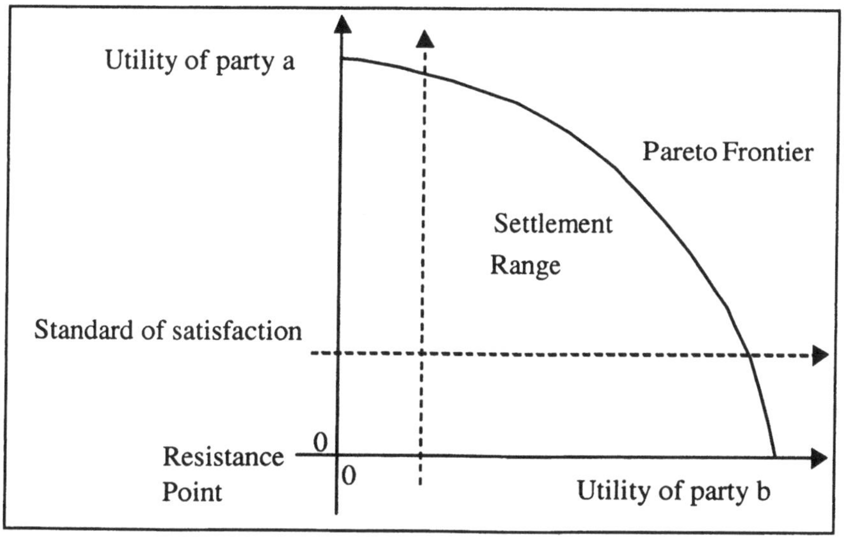

Figure 1. The settlement range. The figure also illustrates how the parties' standards of satisfaction in some cases may restrict the settlement range

According to Zeuthen's theory of "maximum acceptable risk", bargaining situations can be analysed in terms of the relative bargaining power of the two parties (Bacharach and Lawler, 1983). The level of maximum acceptable risk (critical risk) is based on three pieces of information: The current offer of Party A, the current offer of his opponent, Party B, and the conflict point. Each party's critical risk is a function of the ratio of the distance between the current demands of Party A and B and the distance between each party's current demand and a breakdown of the negotiations (Bacharach and Lawler, 1988), as expressed below:

$$\frac{\text{Party's current demand} - \text{Opponent's offer}}{\text{Party's current demand} - \text{Breakdown}} = \text{Critical risk}$$

Critical risk is not, however, an unequivocal guide to behaviour. A party's critical risk is also, for instance, a function of his attitude towards risk and what he considers "acceptable" risk. Thus, in a situation where the parties have identical utilities, and hence, identical bargaining power (a situation which suggests negotiation deadlock), one of them may still make a concession if he is more risk aversive than the other and finds his critical risk less tolerable than the other.

Perhaps even more importantly, given that the situation is characterised by imperfect information, the utilities each party ascribes to his or her own demand, his or her opponent's demand and breakdown are not objective values: They are based on each party's perceptions and subjective evaluations of their own as well as their opponent's utility structure. The accuracy of these perceptions and subjective evaluations, moreover, may depend upon each party's access to reliable and accurate information. According to Morrow, an informational problem occurs when "the actors are uncertain of the value of the available solutions and can benefit by sharing their knowledge", but where distributional interests "prevent the honest sharing of information" (1994: 388). Asymmetries in the parties' access to reliable information may, for instance, in itself affect the parties' (perceptions of) their own as well as their opponent's critical risk. In a situation characterised by uncertainty and imperfect information, therefore, each party's critical risk becomes susceptible to manipulation, and information and knowledge attain a crucial strategic value.[7]

We may assume that in most cases, parties enter negotiations with limited and imperfect information about their opponents' range of alternatives and utility structures. In some cases, they enter negotiations also with limited information – and limited and casual access to information – about factors that may constitute crucial determinants to own utilities and preferences. We

[7] The role of information is a central topic within game theory. This literature is, however, primarily concerned with the impact of information on the parties' course of action or policy choice. Here, we are primarily concerned with the question of how *negotiation dynamics* affect the parties' attitude towards information and knowledge and under which conditions information and knowledge attains a strategic value to the parties of negotiations.

may further assume that communication may affect both the form and the content of an issue's pay-off structure. With this limited information, parties will engage in a search for information to determine the settlement range for the issue and to develop their own preferences to the perceived possible outcomes. They will also adopt tactics instrumental for influencing their opponent's perception of utilities in order to reduce the risks associated with their own preferred courses of action. In this situation, the parties will thus pursue a dual strategy: to gather and search for reliable and accurate information to develop their own preferences, and to influence their opponents' perceptions in his effort to develop own preferences.

This duality implies that information and knowledge become subject to strategic evaluation in terms of their value as tools for achieving political objectives. In distributive bargaining, where values and interests diverge, actors may well find that their own interests are best served if opponents possess inferior knowledge. In general, actors may want their opponents to have inferior knowledge whenever (a) their own interests or values are in conflict with their opponents', and (b) accurate information is believed to favour one set of values or interests relative to another. In these circumstances, incentives for distorting and/or manipulating information are inherent in the strategic logic of distributive bargaining. Even in a situation of conflict, however, a party will also usually want his opponents to have a thorough understanding of the essential features of their common problem. Realising the risk of being misled by misinformation, however, parties' fear of becoming victim to manipulation may sometimes be exaggerated and lead a party to discount also what is in fact "solid knowledge". The mistrust and scepticism with which parties receive (new) information and knowledge in distributive bargaining may thus also ultimately prevent them from communicating true and solid information to their opponents and represents a serious obstacle in their own search for true and reliable information. The sentiment of mistrust which often characterises distributive bargaining thus has significant implications for negotiating parties' orientation towards scientific input. On the one hand, the risk of being misled by distorted and "contaminated" knowledge serves to increase the demand for "pure" and "objective" knowledge. On the other hand, this increased demand for undistorted knowledge also implies that incentives to distort and manipulate information are powerful.

3.3.2 The Role of Information in Integrative Bargaining

The converse case of distributive bargaining is integrative bargaining. In contrast to distributive bargaining, where the subject matter is issues, the subject matter of integrative bargaining is problems. While issues involve a fixed total objective value that can be allocated between the parties in various shares, problems are "agenda items which contain possibilities for greater or lesser amounts of value which can be made available to the two parties" (Walton and McKersie, 1965/93: 127). In its pure form, a problem would be "an agenda item for which the parties would assign the same preference ordering to all possible outcomes and about which the two parties would be equally concerned" (ibid.:127). Hence, while the interests of the parties tend to be diametrically opposed for an issue, the interests of the parties tend to be identical or completely coincidental for a problem. Integrative bargaining may also prevail in situations where the parties' interests are complementary. Accordingly, while a distributive bargaining process is directed towards conflict resolution, a process of integrative bargaining is directed towards *problem solving*. The purpose of integrative bargaining is to maximise the total joint gain. Also, as pointed out above, distributive issues may in some cases be approached by integrative bargaining tactics (Midgaard, 1976).

It should be emphasised that few items correspond to these pure type definitions of issues and problems. Thus, the models are employed in a somewhat relaxed manner. The distributive bargaining model tends to apply in situations predominantly conflictual. Similarly, the subject matter of negotiations is conceptualised as a problem in situations where "the total payoff is varying in sum in a significant way, even though both parties may not share equally in the joint gain, and indeed one may even suffer minor inconveniences in order to provide substantial gains for the other" (Walton and McKersie, 1965/93: 127–8).

There are at least three types of situations that involve an integrative potential. First, there is an integrative potential in situations including agenda items where "one (or more) possible resolution(s) of the agenda item by itself offers both parties a gain in absolute terms over their respective positions in the status quo; for such a resolution neither party experiences any loss" (Walton and McKersie, 1965/93: 128). If these absolutely integrative solutions to an agenda item exist in an objective sense, the situation is said to be of this type – although this may not be obvious to the parties. The second type of situation which involves an integrative potential is less clear-cut, in the sense that none of the possible solutions imply an improvement of status quo to all parties. It is a situation in which "the many possible solutions represent widely varying sacrifice–benefit ratios, but where one solution involves no more than token sacrifice. Thus, the parties must determine the best of the partially integrative solutions" (Walton and McKersie, 1965/93: 129). Third, there is an integrative potential in situations where the parties' interests are complementary and where solutions that offer all parties a gain may be developed by combining and linking these agenda items (Sebenius, 1984).

An integrative potential can be exploited only if it is first discovered, its nature is explored and it is acted upon by the parties. This requires an open exchange of information and knowledge between negotiating parties. Walton and McKersie have developed a model of this process, which involves three steps (ibid.: 137–139):

1. Identify the problem
2. Search for alternative solutions and their consequences; and
3. Develop a preference ordering of solutions and select a course of action.[8]

The first step involves a gathering and exchange of information about each party's perceptions of the problems in order for the problems to be identified and defined in their essentials. Since an integrative potential can only be exploited if it is first discovered, the manners in which problems are defined will have significant impact on the extent to which the integrative potential is realised in the solutions developed. The model assumes that problems may be redefined during the course of the problem-solving process "either as search fails to generate acceptable solutions or as new information suggests a connection with other problems not originally considered" (ibid.: 139).

Step two in the model involves a search for alternative solutions and their consequences. The model assumes that "alternative courses of action (potential solutions) are not immediately apparent but rather have to be discovered or invented" (ibid.: 139). Similarly, it assumes that the full consequences of the alternative solutions may not be obvious, but have to be inferred from analysis of the facts and information available to the parties. In addition to the importance added to an open exchange of information, "invention and creativity are essential in order for appropriate arrangements to be developed for coping with the problems" (ibid.: 139).

Step three "entails the identification of the larger sum of net utilities possible in the situation" (ibid.: 139). This requires openness in the communication of own perceived basic utilities in order to find combinations whereby the integrative potential can best be exploited in the final solution. It is assumed that the parties *do not* have fixed ideas at the *outset* about either how much to search before selecting the "best" solution, or about what would constitute a minimally acceptable solution and search until that is found. In integrative bargaining, the parties engage in a successive comparison of alternatives. Their (current) definition of what would be *minimally successful* is continuously compared with what is (currently)

[8] These steps are analogous to the phases of negotiations in general. They do not necessarily correspond entirely, however. More importantly, the steps in a process of integrative bargaining entail a specific negotiation dynamic which is *not* common to all negotiations.

considered *feasible*. On this basis, they choose whether to accept the best solution generated thus far, or to continue the search – with or without a redefinition of the problem – for a more acceptable solution (Walton and McKersie, 1965/93). It is important to note that this successive process is assumed to be iterative. At any stage of the process, parties may redefine the problem, engage in an additional search for alternative solutions and their consequences or redefine the criteria employed to evaluate alternative solutions if the process does not lead to a satisfactory outcome or as new information appears.

Information thus plays a crucial role also in integrative bargaining and great emphasis is placed on the fact-finding process. In "pure" integrative bargaining, however, information is not subject to the same kind of strategic evaluation as in distributive bargaining. The strategic evaluations involved in integrative bargaining concern the accuracy of the information rather than its value as a tool for misleading opponents and realising individual interests (usually at the expense of other parties). There are also few incentives to distort and manipulate information to mislead opponents. In order to tap the full integrative potential, it is a condition that all parties have access to the best available information. Trust is therefore a precondition for a successful problem-solving process, although complete harmony is not required. At all stages of integrative bargaining the key word is openness in order to make the process work: "For optimum results in integrative bargaining situations, Party must induce the maximum exchange of information relative to mutual problems, so that the problems are identified and defined in their essentials, the fullest range of alternative solutions explored, and the solutions reached represent the relatively higher joint gains possible in the situation" (Walton and McKersie, 1965/93: ibid.: 144).

Equal access to relevant information, however, also requires that each participant have language and other means and skill of communication adequate to take part in the exchange of information. This implies further that the meanings of words are shared fully. When these conditions are not met, a less adequate definition of the problem results (see, for instance, Morrow, 1994).

As our discussion demonstrates, distributive and integrative bargaining represent each other's negation. Tactics instrumental for the one mode of bargaining may be highly dysfunctional for the other. The role of information changes significantly from distributive to integrative bargaining. While being crucially important in both cases, its strategic role in the parties' pursuit of individual interests in distributive bargaining is reduced in integrative bargaining. In integrative bargaining each party's gain is inextricably linked to the extent to which the full integrative potential of the problem is realised. The realisation of an integrative potential requires that all parties have access to the best available information and that all relevant pieces of information are shared fully. In this perspective, concealing or manipulating knowledge to mislead opponents may be directly dysfunctional for a party's realisation of own interests and preferences.

Issues and problems seldom appear in their pure forms, however. Negotiators usually face situations in which agenda items, singly or in combination, involve a significant element of conflict as well as a considerable potential for integration.

3.3.3 The Dilemmas of Mixed Bargaining

Mixed bargaining is a complex combination of distributive and integrative bargaining where each party has a choice between (a) engaging in a search for outcomes with larger total values and (b) directing his negotiation behaviour towards an outcome with a lower total value but which would provide him with a relatively high individual pay-off (Walton and McKersie, 1965/93: 162). Thus, mixed bargaining can be viewed analytically as a complex game involving choices at two levels or along two dimensions (Walton and McKersie, 1965/93: 163). At the first level, the parties must choose between an integrative, increasing-sum strategy and a distributive strategy treating the situation as a fixed-sum issue. In the latter case negotiators will move straight into the second choice between a soft and a hard bargaining strategy in negotiations over the distribution of shares. In the former case, negotiators will engage in a search for integrative solutions and the development of joint gains. Even in the former case, however, negotiations will reach a point where the total amount of the good is identified. The subject matter is then transformed into fixed-sum issues, and negotiations are at that point by definition distributive, implying a second-level choice between a soft and a hard bargaining strategy. Thus, the party confronts a choice between four complex strategies (Walton and McKersie, 1965/93: 164):

1. An integrative strategy to increase the joint gain followed by a soft strategy in the allocation of shares.
2. A distributive strategy, treating the mixed item as fixed-sum, followed by a hard strategy in the allocation of shares.
3. A distributive strategy followed by a soft strategy in the allocation of shares.
4. An integrative strategy to increase the joint gain, followed by a hard strategy in the allocation of shares.

The first two complex strategies imply that the parties behave as if the mixed item were strictly a problem or strictly an issue, adopting an integrative and a distributive bargaining strategy, respectively. The third complex strategy is not recommendable since it can result in the lowest possible outcome for the party adopting it. The fourth strategy appears to be the most attractive: first seeking to increase the joint gain, then engaging in hard bargaining to increase individual shares. This complex strategy is, however, also the most difficult to implement.

The difficulties implied in this strategy stem from the opposite nature of the two modes of bargaining. It may, for instance, prove difficult to separate the distribution of shares from a search for an increasing joint gain. An integrative approach may imply revealing one's true preference orderings. Hence, the party adopting this mixed strategy risks jeopardising his bargaining position in the hard bargaining over individual shares that is to follow. An associated serious difficulty, moreover, is uncertainty regarding the approach of opponents. A party can never be certain that what appears to be integrative bargaining by his opponent is truly that. If the opponent treats the item as a distributive issue, the risks associated with an initial integrative approach are increased.

Thus, a strategy of starting out by treating the mixed item as a variable-sum problem and engaging in a search for joint gains implies a risk of weakening one's own bargaining position at the point when negotiations over the distribution of individual shares start. On the other hand, a strategy of treating the mixed item as an issue from the very start also implies a significant risk. As emphasised above, at the start of negotiations, negotiators very often do not have the information necessary to develop accurate subjective evaluations about the potential gains and losses associated with the alternative strategies. They may not know, for instance, whether there is a positive settlement range if the item is treated as a fixed-sum issue. Thus, by adopting a strategy of distributive bargaining from the very start, they risk an outcome of no agreement in a situation where there might have been integrative solutions to the problem had they only been discovered. Thus in this situation, too, reliable information becomes a key. Each party needs information not only regarding their opponents'/partners' true approach (distributive or integrative), but also about the nature of the item that is subject to negotiations. Given a party's uncertainty about the bargaining tactics adopted by his opponents/partners, however, he will be ambivalent with regard to his receiption of new information and knowledge as well as the openness with which he himself shares information and knowledge.

While the dilemmas associated with mixed-item bargaining are serious and complex, their apparent insolubility may be modified somewhat by taking into account the incremental and progressive nature of the negotiation process. The experience generated in the negotiation process itself is an important source of information. Hence, the amount of information negotiators have increases during the course of the process. Their ability to make accurate assumptions and handle the dilemmas involved in mixed bargaining may therefore vary depending on which phase of the process the negotiations are in. The parties' perceptions of and approach to negotiation problems may, therefore, also change during the course of the process as a function of experience and learning generated by the process itself. In this regard, the negotiation phase represents an important premise for the manner in which negotiation problems are construed, the level of conflict this generates between negotiating parties and the negotiation dynamics which emerge. Also, the kind of information the parties need to develop an effective bargaining strategy may vary according to which phase of the process the negotiations are in.

3.3.4 What Kind of Information is Needed and When?

Generally, negotiating parties seek at least two different kinds of information:

1. Information and knowledge about the nature of the problem in question, and
2. Information about their opponents'/partners' strategy and payoff.

Both of these types of information are equally important for parties to develop a rational and effective negotiating strategy in both integrative and distributive bargaining. The parties do not, however, necessarily seek both types of information equally intensively at all stages of the process. That is, different types of information may be important at different stages of the process.

Zartman and Berman conceptualise negotiations in terms of three main phases[9]: (i) the diagnostic phase, (ii) the formula phase, and (iii) the detail phase. It should be emphasised that the notion of phases is a conceptual tool. In reality a negotiation process is incremental and evolutionary, and in most cases the transition between phases only becomes apparent in hindsight. Moreover, there can be, and usually is, movement back and forth from the one to the other. As pointed out by Zartman and Berman, backtracking is a constant possibility in negotiations. Zartman and Berman's "formula–detail approach" (1982) is, however, a useful theoretical tool to focus on the

[9] While these three phases may overlap with the three steps whereby an integrative potential is identified, explored and acted upon in integrative bargaining (see above), they are not identical in kind. In particular, the negotiation dynamics that characterise integrative bargaining are not common to all negotiations. In most cases, negotiators are confronted with mixed-item problems, and the negotiation dynamics that characterise the three main phases of the process differ significantly from "pure" integrative bargaining.

manners in which negotiating parties' demand for different kinds of information may shift from one negotiation phase to the next.

The diagnostic phase is a pre-negotiation phase in which behaviour is directed towards the identification of the nature of the problem – including each party's position – in order to make issues negotiable. In this phase, parties need to develop an understanding of the nature of the problem in question and whether it requires co-ordinated action. The diagnostic phase, therefore, is essentially a fact-finding endeavour. Especially in cases where knowledge about the subject matter in question is highly specialised, e.g. as is the case with regard to most environmental problems, parties may need external expertise in order to define and diagnose the problem at hand. How a problem is defined will have implications for what needs to be done and who should do it in order to solve it. There may, therefore, be incentives to deceive and distort information. Even in low-conflict situations it may be crucial to employ or establish independent expert groups that to some extent are de-linked from the political deliberations in order to provide all parties with the best available information regarding the nature of the problem in question and hence reduce options for information distortion.

While fact-finding is predominant in the diagnostic phase, parties also engage in an exploration of the political feasibility of joint solutions – i.e., a determination of whether a positive settlement range exists or can be developed. Since negotiations proper have not yet started, there does not seem to be rationale for the parties to bluff and deceive their opponents with regard to their own position at this stage. Rather, the incentives for a party that needs to persuade his opponents about the utility of a negotiated settlement over an existing situation point towards creativity and innovation with regard both to possible solutions and problem definitions in manners that may serve to increase the total utility. Also, parties do not, at this stage, necessarily seek detailed information about opponents' preferences and payoff functions. On the contrary, as pointed out by Zartman and Berman (1982) and others (see for instance, Young and Osherenko, 1993), "creative ambiguity" may be an essential feature of pre-negotiations: "Too much clarity before a position has been jointly defined between parties may make concessions difficult when negotiations begin, and ambiguity may provide flexibility in a prenegotiatory period" (Zartman and Berman, 1982: 71).

In the diagnostic phase, therefore, the parties' need for information and knowledge about the nature of the problem is predominant. Also, the risks associated with an open information exchange seem to be relatively low. However, since the level of conflict between the parties may be expected to increase during the course of the process, it might be considered necessary in some cases at least to have established the mechanism by which independent expert input is to be provided when negotiations start.

At some point during the diagnostic pre-negotiation phase the parties reach the "Turning Point of Seriousness" (Zartman and Berman, 1982: 87): "the perception by each side that the other is serious about finding a negotiated solution – that is, that the other side is willing to 'lose' a little to 'win' a little rather than win or lose all in a non-negotiated approach". When all parties have reached this point (which may not necessarily occur simultaneously), negotiations enter the next stage in which the parties are determined to find a manner by which to arrive at an agreement. At this stage, there are two main routes to the development of agreement (Zartman and Berman, 1982). First, negotiators may approach the task inductively, by building an agreement piecemeal, primarily through mutual compromise or exchanged concessions. In that case, negotiators embark directly on the detail phase. Second, they may adopt a deductive approach, where they "establish first the general principles, or formula, governing the issues susceptible of solution and then work out the implementing details" (Zartman and Berman, 1982: 91).

In that case, negotiations enter *the formula phase*. If possible, efforts to develop a formula for agreement before embarking upon negotiations on the details is recommended because "a formula or framework of principles helps give structure and coherence to an agreement on details, helps facilitate the search for solutions on component items, and helps create a positive, creative image of negotiation rather than an image of concessions and compromise" (Zartman and Berman, 1982: 93). In some cases, however, parties may pursue the development of a formula and negotiations on details more or less simultaneously.

One main task in the development of a formula is to combine or match knowledge and information about the nature of the problem with information about the preference structures and payoff functions of the parties. The point of this exercise is to establish a framework for the development of solutions that are effective in terms of solving the problems for which they were designed that, at a minimum, are politically feasible, and that, to the extent possible, maximise the total joint gain. In this phase, therefore, the parties seek more detailed information about each other's strategy choice and preference structures. While the development of a formula undoubtedly benefits from the open exchange of information associated with integrative bargaining tactics, the negotiation task is typically a mixed-bargaining situation in which the adoption of "pure" integrative bargaining may be associated with a significant risk.

Combining different kinds of information in the development of a formula, parties also need (further) information and knowledge about the nature of the problem in question. Perhaps even more importantly at this stage of the process, they need guidance on how to utilise this kind of knowledge. In a problem definition, the opportunities offered – for instance, with regard to the development of integrative solutions – are not always apparent to negotiators. Thus, especially in situations where a definition of the problem requires expert input and knowledge that negotiators do not themselves possess, the applicability of this knowledge to negotiators and, accordingly, the adequacy of the formula they develop, may depend upon the extent to which such guidance is provided and the knowledge input takes into account the information demand of negotiators in their specific decision-making situation.

In the formula phase of negotiations, therefore, negotiators need information both about the nature of the problem and their opponents' strategy choice and preference orderings. The applicability of the information they receive on the nature of the problem may, however, depend upon the extent to which it is formulated in manners which take the specific decision situation of negotiators into account. The formula phase is typically a mixed-bargaining situation, and the adoption of a "pure" integrative bargaining strategy may be associated with a significant risk. The development of a formula, however, would benefit from an open information exchange between the parties, especially to explore the potential for integrative solutions. A formula provides negotiators with a framework of guidelines and referents for the solution of specific problems. When a formula has been established, either by discovery or invention, negotiations on the details of the agreement can begin (Zartman and Berman, 1982).

The detail phase seems to be predominated by distributive bargaining tactics: "Formulation is difficult because of the innovation required, but detailing is difficult because of the hostility encountered" (Zartman and Berman, 1982: 148). At this stage, therefore, negotiators face the challenge of communicating effectively – exchanging information and tailoring an agreement – within a context of mistrust and in some cases outright hostility. The extent to which hostility is a characteristic feature of the endeavour at this stage may, however, also depend upon the adequacy of the formula. In some cases, major conflicts may have been resolved in the development of a formula, which then would significantly facilitate agreement on the details. In other cases, a formula may prove impossible to transform into a detailed agreement, requiring parties to reformulate the problem definition.

At this stage of the process, the parties' demand for information about their opponents' negotiation strategies, positions and preferences is predominant. This stage of the process is often characterised by the dynamics of concessions and compromise with each party determined not to give in too much. The parties are likely to seek information about the nature of the problem only to the extent that such information may assist them in their pursuit of individual interests. Throughout the process, however, parties may also seek information to support positions into which they are already deeply entrenched. In this regard, information about the nature of the problem is not only sought as a guide to action, but also as legitimisation for positions they would have pursued anyway.

3.3.5 In Sum

Information plays a key role in all negotiations. The openness with which information and knowledge are exchanged may, however, vary significantly according to the nature of the negotiation dynamics. In distributive bargaining, information and knowledge attain a crucial strategic value: Information and knowledge become subject to strategic evaluation in terms of their value as tools for achieving political goals. Hence, parties may perceive their realisation of individual interests as inextricably linked to the concealment, manipulation and distortion of information and knowledge. Thus, incentives for distorting and manipulating information are inherent in the strategic logic of distributive bargaining. In contrast, the extent to which parties are able to identify and fully exploit the integrative potential of bargaining problems, and thereby enhance their own individual gain, depends on an open information exchange whereby all parties have access to the best available information. Thus, while information and knowledge play an equally important key role, they are not subject to the same kind of strategic evaluation.

Most real-world negotiations are characterised by mixed-item bargaining. In mixed bargaining, negotiators face difficult dilemmas that stem from the opposite nature of the tactics associated with the two modes of bargaining: The open information exchange associated with efforts to enhance the total joint gain may weaken the parties' position in the subsequent bargaining over the distribution of shares. Moreover, parties may be uncertain about whether their opponents pursue a distributive or an integrative bargaining strategy. In mixed bargaining, therefore, openness in the exchange of information is associated with a significant risk.

The parties' demand for different kinds of information may vary according to which phase of the process the negotiations are in. In the diagnostic, pre-negotiation phase a demand for information and knowledge about the nature of the problem in question is likely to be predominant, although parties also explore whether a settlement range exists or can be developed. As negotiations proceed to the formula and detail phases, the parties need increasingly detailed information about opponents' negotiation strategies, payoff functions and preferences. The development of a formula requires that knowledge about the nature of the problem is combined and "matched" with information about the parties' payoff functions. In this endeavour, parties may seek guidance on the utilisation of (existing) knowledge about the nature of the problem, rather than new pieces of information. In the concluding detail phase, parties are likely to seek information about the nature of the problem only to the extent that such information may assist them in their pursuit of individual interests. Throughout the process, however, parties may seek information and knowledge about the nature of the problem not only as a guide to action, but as a legitimisation for positions into which they are already deeply entrenched.

3.4 The Dynamics of Science–Policy Interaction

While consensus constitutes a well-established criterion in accordance with which the scientific community distinguishes between established knowledge and knowledge claims, there is significant controversy among science philosophers and sociologists about the nature of the process whereby knowledge is produced. Moreover, while consensus characterises knowledge in the "core", scientific dissent is a distinctive feature of "frontier" knowledge. This indicates that the scientific knowledge which is fed into processes of policymaking is not as robust as the traditional view of scientific knowledge might suggest. At the same time, we have seen that information and knowledge play a crucial role in processes of negotiation whereby policies are developed. Particularly in processes characterised by the logic of distributive bargaining, knowledge may become subject to strategic evaluations in terms of its value as a tool for pursuing political objectives. Incentives to distort and manipulate information and knowledge are thus inherent in the strategic logic of distributive bargaining. The indeterminate nature of scientific knowledge makes it particularly susceptible to such tactics. How does the combination of these features associated with science and politics in their pure forms affect the nature of a process in which they interact? What kind of dynamics are generated in this interaction?

The notion of linearity is a central component of the popular image of the relationship between science and politics. It is assumed that scientific knowledge generated by scientists in accordance with the scientific method at universities and other scientific institutions is merely communicated, undistorted, to policymakers in cases where policymakers need that kind of information as a basis for their (subsequent) policy choice. Thus, the distinction between science and politics is portrayed as clear-cut: science is science, politics is politics and the two (should) never blend. If they do, moreover, the "purity" of the science and, accordingly, scientists' advice is questioned. This image, therefore, also involves a normative dimension according to which science risks political "contamination" and a serious loss of legitimacy in its interaction with politics.[10]

Experience from science–policy interaction processes, for instance in the field of international environmental management, tells us, however, that reality is more complex (see for instance Andresen et al., in press). While science and politics constitute two distinct systems of behaviour, the transformation of scientific knowledge into decision premises for policy choice takes place in a process where these systems of behaviour interact. Hence, a process of science–policy interaction can be perceived of neither as a purely scientific nor as a purely political process. While bearing elements of both science and politics, the nature of processes in which science and politics interact emerges from the interaction itself – when science and politics, each with its own constitutive norms and distinctive features, meet. Only by increasing our understanding of the unique dynamics of science–policy interaction processes may we hope to break new ground on the question of how to achieve a more effective science–policy dialogue in international environmental affairs. That is the task we embark upon in this section.

3.4.1 Scientific Knowledge as Legitimisation for Policy Choice

The image of science as objective and value-free, de-linked from social and political controversy has a strong position in the public, among practising scientists and among policymakers. This image of science gives rise to the belief, first, that nature is the main arbiter of scientific dispute, and second, that scientific consensus, thus developed, will generate political consensus. Thus, science has acquired a very powerful cultural role as a key source of legitimisation for policy choice (Litfin, 1994). The assumption that scientific consensus leads to political consensus has, however, been convincingly demonstrated to be false by several scholars of science–policy studies (*inter alia*, Collingridge and Reeve, 1986; Mulkay, 1978; Jasanoff, 1990). First, it has been argued that consensus in science with policy relevance is argued to be very difficult, if not impossible. Collingridge and Reeve, for instance, maintain that

> relevance to policy, by itself, is sufficient to completely destroy the delicate mechanisms by which scientists normally ensure that their work leads to agreement. Consensus on scientific questions which are more

[10] Another potential risk associated with science–policy interaction is that policy issues are unduly "technified" in the sense that choices of value are disguised as matters of fact. This tendency is particularly associated with the powerful legitimising function of science for policy arguments. This is discussed in more detail below (see Boehmer-Christiansen, 1994).

> than marginally relevant to policy is therefore impossible. Science under these conditions leads not to agreement, but to endless technical bickering about an ever growing number of issues. (1986:ix-x)

Second, it has been shown that even if scientific communities manage to come to some sort of agreement, this will still not necessarily lead to consensus along the political dimensions of the issue in question. Jasanoff, for instance, maintains that "advances in scientific knowledge do not predictably correlate with reductions or increases in policy conflict" (1990:8). Similarly, focusing on the legitimising function of scientific knowledge for policy choice, Litfin directs our attention to the effect this may have on the dynamics of policy debates: each new policy argument introduced with a scientific justification, will lead to efforts on the part of opponents to delegitimise the scientific justification of the argument. She writes:

> the cultural role of science as a key source of legitimation means that political debates are framed in scientific terms; questions of value become reframed as questions of fact, with each confrontation leading to the search for further scientific justification. Paradoxically, the demand for legitimation results in a process of delegitimation. (1994:4)

An explanation and a proposed resolution to this problem has been suggested to lie in the distinction between core and frontier knowledge introduced by Cole (1992). It has been argued that when looking at the role of science in policymaking contexts, one has to take into account the distinction between "facts" and "hypotheses".[11] According to this view, science can only play a constructive role in policymaking when it consists of established facts as opposed to controversial hypotheses. Controversies in science and the inability of science to generate political consensus in policymaking contexts, as discussed by Collingridge and Reeve and Jasanoff may, therefore, be explained simply by the fact that one is trying to "apply" knowledge which *not yet* has attained the status of core knowledge – that is, knowledge which merely consists of (controversial) hypotheses or frontier knowledge. If one were willing to wait for the arrival of consensus in the scientific community, this problem would not arise. This argument, therefore, rests on the assumption that there is a distinction between "basic" and "applied" science. Knowledge, or scientific consensus, is assumed to be established through basic science. Only after this establishment, can knowledge be "applied" in a policymaking context. Moreover, the

[11] This argument was, for instance, put forward at the Conference "Governing Science" held in Oslo, 17. – 18. November 1995.

application of knowledge is not in itself assumed to contribute to the production of new knowledge.

This line of argument, however, does not confront the real dilemma of the situation: To what extent can "science for policy" be anything *but* frontier knowledge? First, in many, if not most cases, if policymakers were to "wait for" the arrival of consensus in the scientific community, it would be too late to solve the problem with which they were confronted. Politics, then, would merely be *re*active rather than *pro*active. Second, and more importantly, science used in a policymaking context, no matter how "consensual," would be taken out of its original context and placed within a new framework – the framework of a specific problem situation not considered in its making. Science for policy will always be an interpretation of a knowledge base, "core" or "frontier". The nature of science implies that it has a certain indeterminacy attached to it. Very few pieces of (core) knowledge are readily available for policymakers to "apply", or are able to give unambiguous answers to what to do in a specific situation to mitigate or avoid specific problems. Mulkay, for instance, has found that "members of the same specialized and mature research community frequently reach different conclusions when they try to apply their expertise in practical situations" (1978:118). This indicates that scientific consensus may not only be context-dependent, but also that "intellectual consensus in science is relatively loose and flexible, and that its content is open to interpretation in numerous directions" (Mulkay, 1978: 118). In order to be useful as premises for policymaking, therefore, knowledge has to be *transformed* into policy advice (see also Litfin, 1994; Jasanoff, 1990; Sundquist, 1978). This implies that scientific knowledge in a policymaking context will always have the character of frontier knowledge, with the vulnerability this implies towards efforts of manipulation and distortion.

The combination of these features – the legitimising power of science in policymaking and the indeterminacy of scientific knowledge, especially in combination with political conflict – may give rise to a rather perverse dynamic in processes of science–policy interaction. On the one hand, policymakers will be engaged in a strenuous attempt to find a manner in which to couch a policy argument in scientific terms and to delegitimise the scientific justification of the arguments of their opponents. On the other hand, given the indeterminate nature of scientific knowledge, science is very susceptible to this kind of manipulation. Scientific communities may become involved in disputes of method and interpretation ("endless technical bickering") often not realising the political role of such disputes in processes of policymaking. Or, when realising the political role of such disputes, they may become tempted to conceal points of contention within the scientific community and exclude (from the science–policy interaction process) members of the community who hold opposing views, thus acquiring roles as advocates rather than scientists. Although a basis of core knowledge to some extent may place constraints on the free play of such mechanisms, the existence of core knowledge cannot entirely prevent their operation in the process.

It may be assumed that the operation of this vicious dynamic is correlated with the nature of the subject matter over which parties negotiate and the level of political conflict between parties in pursuing their interests. Situations pre-dominated by perceptions of fixed-sum issues and a high level of political conflict increases each party's sensitivity to the risk of falling victim to information distortion, manipulation and deception. The sentiment of mistrust thus emerging may serve to increase policymakers' scepticism towards scientific input, as well as the strength by which efforts to manipulate the scientific knowledge base are pursued. Paradoxically, since persuasion may play an important role in distributive bargaining, the political power of scientific facts also increases. In negotiations dominated by distributive bargaining behaviour, the mechanisms inherent in science–policy interaction are thus reinforced. While a predominance of integrative bargaining behaviour may restrict the operation of such mechanisms somewhat, negotiators' constant awareness of efforts at manipulation will serve not to remove them altogether.

As the preceding discussion demonstrates, reliable information, not only regarding each party's position but also regarding the nature of the problem in question – the facts of the matter – plays a crucial role in negotiations. In order to assess the reliability of a piece of information, negotiators need to know how that piece of information has been brought about. First and foremost, they are likely to consider the scientific integrity they associate with the scientists or scientific institutions who provide the information. As will be further discussed in the next chapter, scientific integrity may be regarded as a function of a set of factors, but as a general rule we may assume that negotiators will be more sceptical towards information brought about by "foreign" scientists than scientists from their own country (see, for instance, Andresen and Østreng, 1989). Moreover, they are likely to be less sceptical towards information brought about by institutions recognised as "scientific" than towards information brought about by institutions that may have even the remotest interest in the subject matter at hand. On the other hand, however, in order to get the particular points of information they need in their specific situation, they would also probably prefer a relatively close relationship with the scientists and scientific institutions that provide the information. If they are positively certain that the scientists involved in this dialogue are not biased, at least not in direction of the interests of their opponents, this kind of involvement by scientists in the policymaking process would not be problematic. Since all parties to negotiations may be assumed to have these preferences, however, there is a problem. When each party engage "their own" scientists, the dynamics of legitimisation–delegitimisation of policy arguments takes effect.

3.4.2 "Negotiating" Knowledge

One way this potentially paralysing mechanism inherent in science–policy interaction may be avoided, or at least restricted, is to establish institutional structures within which an understanding and interpretation of the problem within its political context may be developed. In this way, scientists and policymakers can in concert develop a "new" consensus defining the characteristic features of the problem in question within its political context. The provision of arenas permitting a high level of integration and involvement between science and policy thus seems to be required to tackle this potentially paralysing pit-fall of science–policy interaction. On the other hand, this process operates in the shadow of the public image of science, according to which this kind of involvement in itself could imply a serious loss of scientific legitimacy. Also, policymakers themselves are likely to have more faith in conclusions generated in a process which is unambiguously "scientific" in nature. This implies that the process is most likely to fail unless the institutional structures within which the science–policy dialogue takes place are also separated from the political sphere and acknowledged – by scientists, policymakers as well as the lay public – as "scientific". Our discussion thus generates the proposition that science–policy interaction processes are most likely to succeed if they are organised so as to provide arenas for a close dialogue, among and especially between scientists and policymakers, within the framework of institutions acknowledged as "scientific".

In her studies of regulatory or advisory science in the US, Sheila Jasanoff found that "negotiation – among scientists as well as between scientists and the lay public – is one of the keys to the success of the advisory process" (1990: 234). She further suggests that this indicates that a strict separation between science and politics, in this context, is artificial:

> The negotiated and constructed model of scientific knowledge, which closely captures the realities of regulatory science, rules out the possibility of drawing sharp boundaries between facts and values or claims and context. ... Evidence from regulatory case histories suggest, further, that proceedings founded on the separatist principle frequently generate more conflict than those which seek, however imperfectly, to integrate scientific and political decisionmaking. (ibid.: 231)

The paradox is, however, that although this close dialogue constitutes an essential part of a successful scientific advisory process, an equally essential part of the process, according to the studies of Jasanoff, is to draw seemingly sharp boundaries between science and policy. These boundaries are not, however, regarded as objective features of science, but are regarded as socially negotiated through the micro-process of "boundary work". While negotiation is seen as an important device for providing an outcome acceptable to all participating actors (scientists as well as policymakers), "boundary work" is the process that gives this outcome legitimacy:

> By drawing seemingly sharp boundaries between science and policy, scientists in effect post 'keep out' signs to prevent nonscientists from challenging or reinterpreting claims labeled as 'science.' The creation of such boundaries seems crucial to the political acceptability of advice... Curiously, however, the most politically successful examples of boundary work are those that leave some room for agencies and their advisers to negotiate the location and meaning of the boundaries. (ibid.: 236)

With regard to the nature of processes of science–policy interaction, Jasanoff concludes that:

> What emerges from a successful recourse to scientific advice, then, is a very special kind of construct: one that many, perhaps most, observers accept as science, although it both shapes and is shaped by policy. ... When the process works, few incentives remain for political adversaries to deconstruct the results or to attack them as bad science. (ibid.: 234-7)

It is also interesting to note that Gibbons et. al. (1994) find evidence of a more general trend towards a new mode of knowledge production, developed in the context of application:

> A number of attributes have been identified which suggest that the way in which knowledge is being produced is beginning to change. To the extent that these attributes occur across a wide range of scientific and scholarly activity, and persist through time they may be said to constitute trends in the way knowledge is produced. ... [It appears] that they occur most frequently in those areas which currently define the frontier and among those who are regarded as leaders in their various fields. ... [T]hese trends, because they seem to involve the intellectual leaders, probably ought not to be ignored. (p.1)

The knowledge produced according to this new mode has certain characteristic features:

Such knowledge is intended to be useful to someone whether in industry or government, or society more generally and this imperative is present from the beginning. Knowledge is always produced under an aspect of continuous negotiation and it will not be produced unless and until the interests of the various actors are included. Such is the context of application" (Gibbons et. al., p. 4).

3.5 In Sum

This analysis indicates that while the scientific process may essentially be seen as a truth-seeking endeavour, and consensus constitutes the main criterion whereby established knowledge is distinguished from mere knowledge claims, it is also a social process where the road to consensus is not as straightforward as the traditional image of science might suggest. Thus, policymakers may be confronted by scientific dissent which cannot be dismissed as value bias or incompetence on behalf of the scientists.

This analysis also shows that accurate and reliable information is crucial in all negotiations. Information and knowledge may, however, become subject to strategic evaluation in terms of their value as tools for achieving political goals. Thus, incentives to distort and manipulate information are inherent in the strategic logic of distributive bargaining. Even in integrative bargaining, an open information exchange can be associated with a serious risk, at least if parties are uncertain about the true bargaining strategy of their opponents.

Science–policy interaction is thus characterised by an immanent tension between impartiality and disinterestedness on the one hand, and strategic reasoning and interest realisation on the other. Combined with the indeterminacy and context dependency of scientific knowledge and its power as legitimisation for policy arguments, this tension may give rise to a dynamics that may be highly counterproductive in terms of the effectiveness of the science–policy dialogue. Such dynamics, moreover, may be assumed to be positively correlated and hence reinforced by political conflict.

Science–policy interaction may, therefore, especially in situations characterised by a high level of political conflict, be subject to a vicious dynamic of scientific legitimisation and delegitimisation of policy arguments. One manner by which to cope with this potentially paralysing pit-fall inherent in science–policy interaction may be to establish institutional structures providing arenas for a close science–policy dialogue that at the same time are recognised as "scientific". This discussion thus suggests that in order to increase the effectiveness of science–policy interaction processes, institutional structures facilitating the production of new knowledge in the context of its application, characterised by a high level of integration between science and policy while at the same time maintaining the scientific integrity and authority of the knowledge provided, are required. The extent to which and how this may be achieved is subject to discussion in the next chapter.

Chapter 4

Designing Institutions for Science–Policy Interaction

4.1 Introduction

This analysis aims to investigate the extent to which the institutional framework encompassing the science–policy dialogue may serve as an instrument for improving the effectiveness of the endeavour. The immediate source of this focus of study is the increasing interest among negotiation theorists in the manners in which the institutional arrangements of negotiation processes may affect negotiation behaviour by inducing or restricting particular courses of action (see *inter alia* Buzan, 1981; Miles, 1977; Nordquist, 1985; Sanger, 1987; Sebenius, 1984; Underdal, 1989; Winham, 1977; Zartman, 1975 and 1978). The interest in institutional structures as modifying agents to individual actor behaviour originates in organisational studies (see for instance, Egeberg, 1984; Scott, 1981; March and Olsen, 1989; Powell and DiMaggio, 1991), and surfaced within international studies particularly with the intense focus on international regimes from the latter part of the 1970s onwards (see especially Keohane and Nye, 1977; 1987; Krasner, 1983; 1988; Young, 1989). Regime theory emerged as a reaction to the anarchic and state-centred approach to international relations characterised by structural realism (see especially Waltz, 1979; and for a critique of this approach see Keohane (ed.), 1986). In contrast to structural realists, regime theorists claim that states do not operate in an entirely anarchic system in all areas of international relations. While there is no authority at the international level corresponding to nation state governments, actors' behaviour is still, in some areas, restricted by "sets of implicit or explicit principles, norms, rules, and decision-making procedures around which actors' expectations converge in a given area of international relations" (Krasner, 1983) – i.e., international regimes. Hence, states behave as rational utility maximisers within social institutions that also constrain their behaviour.

At the micro level this "new institutionalism" has also found its expression in the assumption that negotiation outcomes cannot be explained entirely as the aggregated outcome of each actor's pursuit of individual interests; the institutional framework within which the outcome has been developed may have independent explanatory power. This has led to an increasing interest in the manners by which institutional arrangements, as social constructs, can be utilised as potential instruments to guide actor behaviour in order to prevent or restrict courses of action that are particularly dysfunctional in terms of developing agreements (see especially Underdal, 1989; 1990; 1991a).

This chapter investigates the extent to which this approach is applicable in the context of a process of science–policy interaction. It addresses three main questions: (i) To what extent can behaviour in science–policy interaction be assumed to be guided by institutional arrangements? (ii) To what extent can institutions for science–policy interaction be designed? The answers to these questions determine the extent to which institutional design may be utilised as an instrument to facilitate and enhance the effectiveness of the science–policy dialogue and pave the way for the third question: (iii) By which institutional devices can this potential be realised?

The second objective in this analysis is to investigate the role that leadership performances may play in processes of science–policy interaction and which leadership functions that may be assumed to be in demand during different parts of this process. In negotiation analyses, leadership behaviour in the negotiation process is increasingly seen as an important factor both in the development of institutions that may facilitate the negotiation process and in the development of (integrative) solutions (see Young, 1991; Underdal, 1991a,b; 1994; Malnes, 1995).

Section 4.2 discusses how institutions are defined, especially with regard to the distinction between institutions and organisations. Questions (i) and (ii) are addressed in section 4.3, question (iii) is addressed in section 4.4, and the question of leadership performance is addressed in section 4.5. Section 4.6 discusses other factors that may explain the extent to which scientific knowledge is acted upon by policymakers.

4.2 Institutions Defined

Institutions may be defined as "sets of rules of the game or codes of conduct that serve to define social practices, assign roles to the participants in these practices, and guide the interactions among occupants of these roles" (Young, 1994: 3). As Young points out, the concepts of "institution" and "organisation" should not be used interchangeably:

> Whereas institutions are sets of rules of the game or codes of conduct defining social practices, organizations are material entities possessing offices, personnel, budgets, equipment, and, more often than not, legal personality. Put another way, organizations are actors in social practices. Institutions affect the behavior of these actors by defining social practices and spelling out codes of conduct appropriate to them, but they are not actors in their own right. (Young, 1994: 4)

In this case, negotiations or processes of science–policy interaction may or may not take place within the framework of organisations. In some cases international organisations may constitute actors in processes of negotiation (for instance the European Union). In many cases negotiations take place under the auspices of one or several international organisations (for instance UN organisations regularly serve this function). Thus, the distinction between institutions and organisations is of relevance in this context. The main issue, however, is what determines the instrumental potential of institutional arrangements, and the question of whether or not these arrangements are embedded in organisational structures is not addressed. In the literature upon which this discussion is based, the distinction between the concepts of "institution" and "organisation" is, however, often less clear-cut. In most cases references in the literature to "organisational structures" are equally valid for the institutional structures of social institutions. In the following, therefore, the two are treated synonymously.

4.3 Do Institutions Matter, and Can They be Designed?

While institutions may vary along a number of dimensions – functional scope, degree of formalisation, geographical domain etc. – all institutions are nonetheless "social artefacts" that are consciously or unconsciously created by human beings (Young, 1994). Although socially constructed, however, they are not always the result of human design. The extent to which institutional arrangements may be employed as tools for guiding actor behaviour and thus improving institutional performance may be assumed to be dependent upon at least two criteria:

1. the extent to which the institutional structures, constituted by the rules guiding behaviour and prescriptions of roles and role relations, are deliberately designed and formalised, and
2. the extent to which actor behaviour is in fact guided by these formalised structures in systematic and predictable ways.

4.3.1 Social Institutions as Rational, Natural and Open Systems

On the basis of organisation theory, we may distinguish between three perspectives of social institutions according to variations along these dimensions. First, social institutions may be described as rational systems: collectivities oriented towards the pursuit of relatively specific goals according to a formalised structure. A structure is formalised to the extent that "the rules governing behaviour are precisely and explicitly formulated and to the extent that roles and role relations are prescribed independently of the personal attributes of individuals occupying positions in the structure" (Scott, 1981: 20). Rational system theorists emphasise the importance of goal specificity and formalisation "because each of these elements makes an important contribution to the rationality of organizational action" (Scott, 1981: 58). Specific goals are needed not only for rational assessment and choice to be possible; they also guide decisions about how the institutional structure itself is to be designed. Hence, this perspective assumes that institutions are the result of human design and that actor behaviour is actually guided by this formalised structure. Social institutions defined as rational systems are thus the archetype of institutions within which institutional design can be employed as a tool for guiding behaviour and improving institutional performance: "in a fundamental sense, the organizational structure is viewed as a means, as an instrument, which can be modified as necessary to improve performance" (Scott, 1981: 60).

Second, institutions may be described as natural systems. Whereas rational systems are seen as collectivities constructed to pursue specific goals, natural systems are conceived of first and foremost as collectivities (Scott, 1981). According to this perspective the institution is seen as "a self-maintaining system that must satisfy a stable set of internal needs at the same time that it must adapt to influences impinging on it from an external environment" (Scott, 1981: 80). Institutional energy is thus given to self-maintenance goals. Institutions defined as natural systems may or may not be characterised by goal specificity and a formalised structure. In cases where the institution is characterised by these features, a discrepancy between the formalised structures and actual actor behaviour may occur in situations where the specified goals of the institution are regarded as incompatible with the overriding goal of survival. Thus, within institutions that correspond to this natural perspective, informal norms and rules may, in given situations, take precedent over the formal rules as guides to behaviour. Moreover, whereas rational systems are designed, natural systems evolve: "the former develop by conscious design, the latter by natural growth; rational systems are characterized by calculation; natural by spontaneity" (Scott, 1981: 101).

Finally, institutions may be described as open or loosely coupled systems, constituted by "a coalition of shifting interest groups that develop goals by negotiation; the structure of the coalition, its activities, and its outcomes are strongly influenced by environmental factors" (Scott, 1981: 23). Institutions defined as loosely coupled systems are characterised by a low degree of goal specificity and a low degree of formalisation. Actor behaviour is thus not, or to a very little extent, guided by any formal structure. Rather, behaviour is seen as the result of individual rational choice based on calculations of costs and benefits associated with particular decision situations (Egeberg, 1984). In open or loosely coupled systems it is, therefore, precisely the lack of constraints that guides behaviour. Examples of social institutions that work according to this definition may be found especially at the international level where certain types of international regimes may be found to correspond this definition. Initially international regimes were assumed to emerge to the extent that nation states, as rational, self-interested utility maximisers regarded them as beneficial according to some long-term interest: "If no benefits were realized from international agreements or if cooperation could be sustained without cost, international regimes would not arise" (Powell and DiMaggio, 1991: 7). Moreover, changes in coalitions of interests constituting the basis for international regimes are also assumed to bring about regime change. This rationalistic approach to the study of international regimes has, however, been questioned (see for instance Keohane, 1989, chap. 7). Actor behaviour in political elections has also been sought to be explained in these terms, although such efforts have been criticised for their lack of empirical accuracy (see for instance Lewin, 1988). These criticisms notwithstanding, institutions operating in accordance with an open systems definition would hold the least potential as a tool for guiding or modifying behaviour, their low degree of formalisation taken into account.

4.3.2 Institutions as Instruments

Social institutions are characterised by varying levels of formal institutionalisation. Institutions corresponding to the rational systems definition represent the most formalised while institutions corresponding to the open systems definition are placed at the other end of this continuum representing the least formalised. Formal institutionalisation is a condition for designability. Moreover, in cases where informal rules and norms take precedent over a formalised structure as guides to behaviour, options for employing formal institutional arrangements as instruments for improving institutional performance are reduced accordingly. Where may we assume institutions for science–policy interaction to be placed along this continuum?

As emphasised in chapter 3, processes of science–policy interaction are first and foremost characterised by the interaction between two distinct systems of behaviour, each with its own constitutive norms, rules and procedures, formally or informally established, according to which behaviour within each system – in their pure form – is guided. Whereas science as institution may be assumed to resemble a natural system, international politics in its pure form may be placed closer to the open system perspective. In their pure forms, therefore, each of these systems of behaviour may seem to leave little room for institutional design as instrument.

In cases where formal institutions for science–policy interaction have not been established, we may further assume that the science–policy dialogue will be particularly vulnerable to the legitimisation–delegitimisation dynamics discussed in chapter 3. The vulnerability of the process to these mechanisms may also be assumed to be positively correlated with the level of political conflict characterising the issue area; i.e., a high level of political conflict is followed by increased vulnerability to an endless process of legitimisation–delegitimisation of policy arguments. This point will be further elaborated below.

If, on the other hand, formalised institutional structures are developed within which the science–policy interaction process takes place, this formalisation in itself enhances the potential of institutional arrangements as instruments for improving the effectiveness of the science–policy dialogue, but only up to a certain point. Recalling the discussion of the natural-systems perspective of social institutions, it may be assumed that science as an institution will be guided by the formal institutional structures of the endeavour only to the extent that the specified goals and procedures by which to achieve these goals are compatible with scientists' perceptions of the overriding goal of survival. Survival in this case may be assumed to be associated with maintenance of scientific integrity and autonomy. This suggests that formalised institutional structures will guide behaviour in processes of science–policy interaction only to the extent that they are compatible with the overriding goal of maintenance of the scientific integrity of scientists taking part in the process. It can thus be assumed that in cases where scientists experience a discrepancy between the specified goals, or the procedures by which to achieve these goals, and their overriding concern for maintenance of scientific integrity, that informal norms, rules and procedures, notably those guiding science at large, will take precedent over the formalised institutional structure as guides to behaviour.

Political conflict may also be assumed to affect the extent to which actor behaviour is guided by formalised institutional arrangements. Institutional design is also a negotiated outcome and political conflict may penetrate and contaminate discussions of institutional arrangements and thereby restrict the flexibility and availability of this instrument. It can thus be assumed that it is more difficult to develop agreement on institutional arrangements in distributive than in integrative bargaining. Moreover, institutional arrangements may also, particularly during the initial stages of negotiations, affect the extent to which problems are perceived as distributive in the first place.

With regard to processes of science–policy interaction, on the other hand, we may assume that the level of political conflict may affect the vulnerability of the process to the legitimisation–delegitimisation dynamics. When the level of political conflict is low, the process is less vulnerable to this mechanism than when the level of political conflict is high. With regard to processes of science–policy interaction, therefore, we may assume that while a high level of political conflict may serve to restrict the availability of institutions as instruments, the parties' need for institutional arrangements capable of restricting the functioning of the legitimisation–delegitimisation mechanism of policy arguments increases.

4.4 Realising the Instrumental Potential of Institutional Design

This discussion suggests that the institutional framework within which processes of science–policy interactions take place has an instrumental potential under certain conditions – notably in cases where the institutional structures are formalised – and within certain limits, notably to the extent that they are compatible with an overriding goal of maintenance of the scientific integrity of participating scientists and within the restrictions caused by political conflict. How can this instrumental potential be realised in order to enhance the effectiveness of the science–policy dialogue?

4.4.1 Institutional Functions

This question is approached by focusing on which functions the institutional framework of the science–policy dialogue should be able to serve in order to enhance its effectiveness, and to which institutional arrangements these functions are assumed to be linked: Towards which goals is institutional design a potential means and by which institutional devices?

As discussed in chapter 2, the effectiveness of processes of science–policy interaction can be judged against two measures: the extent to which policymakers have accepted the factual validity of the knowledge base, and the extent to which they have also accepted its policy implications and acted upon the knowledge provided. Policymakers' acceptance is conceptualised in terms of the development of a consensual problem diagnosis (see chapter 2).

It is assumed that the institutional arrangements for the science–policy dialogue have an indirect impact on the extent to which policymakers choose to act upon the knowledge provided: The institutional arrangements of the dialogue are assumed mainly to affect policy choice via policymakers' acceptance of the knowledge base. This does not imply, however, that the development of a complete problem diagnosis is required for policymakers to act. Recalling our discussion in chapter 2, a consensual problem diagnosis is conceptualised in terms of two constitutive elements: (i) a knowledge base, whose representativeness of state-of-the-art knowledge is acknowledged by scientists and policymakers, and (ii) a consensus on the linkage between this knowledge base and valued policy goals. It is only the first of these elements that is considered a necessary condition for policymakers' to choose to act upon the (policy implications of the) knowledge provided.

Thus, institutions for science–policy interaction should be designed to facilitate the development of a consensual problem diagnosis in terms of both of its constitutive elements. While policymakers may chose to act in the absence of a (complete) problem diagnosis, it is assumed here that the development of a consensual problem diagnosis facilitates the process through which policies are developed and decided upon.

Policymakers' acceptance of the knowledge base and the development of a problem diagnosis may be assumed to be linked to the extent to which:

a) the knowledge base is provided by scientists whose scientific authority and integrity they acknowledge;
b) channels for communication and dialogue between scientists and policymakers are established;
c) all concerned groups and interests are represented in the scientific bodies, which is operationalised as the extent to which the knowledge base is provided by a group of scientists/experts with a geographic composition which is representative of the composition of parties in the policymaking processes in the issue area;
d) the knowledge base is relevant and applicable to the issues and issue areas of particular interest to policymakers in their decision-making situation;
e) arenas for communication, interactive dialogue and, if necessary, negotiations among and especially between scientists and policymakers are established; and
f) formal and/or informal mechanisms are developed whereby possible points of conflict or strong disagreements among and between scientists and policymakers may be handled and resolved.

These six factors may be summarised into four main institutional functions the framework should be able to serve in order to enhance the effectiveness of the science–policy dialogue:

1. *Scientific autonomy and integrity*: In order to increase the likelihood of policymakers' acceptance and adoption of the knowledge base and in order to maintain scientists' overriding goal of scientific autonomy, the institutional framework should be designed so as to maintain the scientific autonomy and integrity of the scientists and scientific bodies in the process. In order to serve this function, institutional devices that serve to separate science and politics are needed.
2. *Science–policy integration*: In order to establish arenas for an interactive dialogue between scientists and policymakers, some level of involvement between science and politics – whereby scientists to some extent become involved in the policymaking process and policymakers to some extent become involved in the process by which a knowledge base is developed – is required. Such arenas would also serve to improve communication channels between scientists and policymakers and thereby increase the likelihood that the scientific knowledge base provided is relevant and applicable to policymakers in their decision situation. In order to serve this function, institutional devices that serve to integrate science and politics are needed.
3. *Geographical representativeness*: When processes of science–policy interaction are linked to policymaking at the international level, particularly in cases where the North–South controversy is mobilised, it is important that the geographic composition of the group of scientists/experts is representative of the geographic composition of parties to the policymaking process, at least in terms of geographic region. In order to serve this function institutional devices that ensure a fair geographic balance in the group of scientists are needed.
4. *Conflict resolution*: In the development of a consensual problem diagnosis, conflicts or points of disagreement may arise. It is therefore important that the institutional framework provides mechanisms whereby conflicts or disagreements can be handled. In order to serve this function, institutional devices for conflict resolution are needed.

4.4.2 Linking Functions to Institutional Devices

Our preceding discussion suggests that in a seemingly paradoxical sense the effectiveness of institutions for science–policy interaction seems to depend upon, *inter alia*, the extent to which they are capable of ensuring both a separation and a certain level of integration of science and politics. To what extent is it at all possible to develop institutions where these functions are combined?

The extent to which these intuitively contradictory functions can be combined depends upon the extent to which they are linked to the same institutional devices. If these functions are linked to different institutional devices, the maintenance of both is, at least theoretically, possible. The following section, therefore, explores which institutional devices these functions can be assumed to be linked to, and then returns to the question of their compatibility.[1]

Maintenance of the scientific autonomy and integrity of the scientists and scientific bodies of the interaction can be conceived of as a function of multiple variables. The main factors in this regard can be divided into the following six dimensions:

1. *Appointment of scientists*
   - Non-governmental, scientific organisations, independent of a regulatory body
   - National governments
   - Intergovernmental organisations
   - A regulatory body

2. *Funding of scientists*
   - Non-governmental, scientific organisations, independent of a regulatory body
   - National governments; or allocation of funds from national governments via regulatory body
   - Intergovernmental organisations

3. *Principal criteria for recruitment*
   - Entry open to any participant nominated by a member country
   - Entry subject to restrictions
   - To the extent that entry is subject to restriction, a principal criterion is scholarly merit

4. *Main function of the body*
   - Production of new knowledge; participants active in research
   - Co-ordination of research

[1] Particularly with regard to institutional devices instrumental for maintaining scientific autonomy/integrity and ensuring some level of involvement between science and politics, this discussion draws heavily upon Andresen et al. (in press).

- Interpretation of existing knowledge, particularly with regard to elaboration of policy implications in a particular policymaking context
- Policy advice

5. *Operational autonomy*
   - The extent to which the group of scientists is free to organise its own work: Can the group set its own agenda, decide on division of labour and the allocation of specific roles?

6. *Unity of scientific body/network*
   - Does the scientific body or network have its own institutional basis, i.e., some internal co-ordinating device?
   - Are all members scientists, or are other professions (administrators or other non-scientists) also included?

By combining these dimensions an index may be constructed which suggests the relationship between these institutional devices and the maintenance of scientific autonomy/integrity, as indicated in table 4.1.

Table 1. Index of the autonomy/integrity of the scientific bodies of the interaction

High autonomy/integrity	**Low autonomy/integrity**
Appointment by scientific organisations or IGO	Appointment by national governments or regulatory body
Funding by scientific organisations, regular channels for scientific funding at national level, or IGO	Funding by national governments, or allocation of funds from national governments via regulatory body
Recruitment based on scholarly merit or role in scientific community	Recruitment based on political or administrative position
Participants engaged in active research or co-ordination of research	Participants mainly concerned with policy implications or policy advice
High operational autonomy: sets its own agenda and organises its own work	Low operational autonomy: under instruction and control by governments

High autonomy/integrity	Low autonomy/integrity
Independent institutional basis: internal co-ordination, all members are scientists	No independent institutional basis, includes also other professions than scientists

Similarly, the level of involvement between the scientific and the political bodies of the interaction is conceived of as a function of the extent to which scientific bodies are engaged in formulating policy advice (dimension 4 above) and the following three additional dimensions:

7. *Functional differentiation*
   - To the extent that the scientific bodies provide policy advice, are the functions of research/research co-ordination and formulation of policy advice separated, or do the same individuals or groups of individuals perform both functions?
   - To the extent that these functions are separated, how closely linked are the bodies performing the functions?

8. *Formal links to decision-making body*
   - To what extent are there regular channels for communication between scientific and political bodies?

9. *The provision of arenas for interactive dialogue*
   - To what extent are there arenas for an interactive dialogue between scientists and policymakers that provide opportunities for simultaneous discussions among and between scientists and policymakers on interpretations of scientific knowledge and their policy implications?
   - To the extent that such arenas are provided, to what extent are they institutionally separated from bodies conducting research/research co-ordination and the regulatory body?

Table 4.2 indicates the assumed relationship between these factors within each dimension and the level of integration between the scientific and political bodies of the interaction. The table also illustrates the assumption that arenas for an interactive dialogue may be provided in manners representing a "medium" level of integration.

Representativeness is primarily linked to the rules of procedure for participation. The geographic representativeness of scientific bodies can be assumed to be high to the extent that rules ensuring a geographic balance actually guide the composition of participants in these bodies.

Table 2. Index of level of involvement between the scientific and political bodies of the interaction.

<table>
<tr><th>High level of integration</th><th>Low level of integration</th></tr>
<tr><td>The functions of research/research co-ordination and formulation of policy advice are not separated, scientists formulate policy advice, or</td><td>The functions of research/research co-ordination are separated; scientists do not formulate policy advice</td></tr>
<tr><td>Alternatively: The functions of research/research co-ordination and formulation of policy advice are separated, but the bodies serving the functions are closely linked</td><td>The functions of research/research co-ordination are separated, and the bodies serving the functions are not closely linked</td></tr>
<tr><td>There are regular channels for communication between scientific and political bodies</td><td>There are no or only informal channels for communication between scientific and political bodies.</td></tr>
<tr><td>Arenas for an interactive dialogue among and between scientists and policymakers are provided which are not clearly institutionally separated from bodies conducting research/research co-ordination and regulatory body.</td><td>There are no arenas within the formal framework for an interactive dialogue among and between scientists and policymakers</td></tr>
<tr><td colspan="2">There are arenas for an interactive dialogue among and between scientists and policymakers, but they are clearly separated from bodies conducting research/research co-ordination and regulatory body. Alternatively: there are informal arenas for an interactive dialogue among and between scientists and policymakers, and they are thus clearly separated from bodies conducting research/research co-ordination and regulatory body.</td></tr>
</table>

To determine which institutional devices that may be of importance in the establishment of mechanisms for conflict resolution in processes of science–policy interaction, we need to take one step back and explore what kind of conflicts that can be anticipated in the development of a consensual problem diagnosis.

In negotiations, conflicts are very often linked to distributive issues. In the context of a science–policy interaction process, conflicts may be assumed to take the form of

- disagreements over the interpretation, framing and hence presentation of scientific knowledge,
- disagreements over the appropriate borders between science and politics, and
- disagreements linked to the elaboration of policy implications.

While the establishment of arenas for an interactive dialogue between scientists and policymakers may serve to bring such disputes into light, they may also constitute a means by which to handle them, simply by serving as arenas where such issues can legitimately be brought up and discussed. One institutional device by which to establish mechanisms to handle disputes is therefore linked to the level of integration of science and politics, particularly the extent to which arenas for an interactive dialogue are provided (dimension 9 above).

The disputes that may be associated with processes of science–policy interaction, however, may be, and often are, indirectly linked to distributive issues. Even so, within this context, straightforward negotiations can be assumed to be an illegitimate procedure for handling such situations, since that would suggest that the knowledge base is a negotiated outcome and thus seriously jeopardise the scientific authority and credibility of the knowledge base. Negotiation theory, therefore, can provide only limited guidance as to how such situations can be handled. One main recommendation from negotiation theory for handling issues of dispute, the transfer of the discussions to informal arenas, may, however, be useful in this context. Especially since some of the parties to such conflicts may have hidden agendas (for instance, expected distributive implications from alternative presentations of scientific knowledge), it may be important to move deliberations to arenas beyond the limelight of public attention and get the real issues on the table. On the other hand, transparency and openness are also features of importance for the success of a science–policy dialogue, and these may be incompatible with the employment of side-meetings and other informal arrangements as a means for handling situations of dispute. Employed with care and "diplomatic finesse", however, the establishment of informal arenas may constitute a tool for dealing with disputes in processes of science–policy interaction.

In disputes over the interpretation and presentation of scientific knowledge, moreover, we may assume that a decision procedure based on positions of authority – where, more specifically, scientists are recognised as "authoritative interpreters" in the sense that scientists have a final say in the matter – can to some extent be employed. This mode of decision-making is assumed to be effective as a mechanism for handling disputes to the extent that policymakers acknowledge the scientific authority of the scientists.

Thus mechanisms for conflict resolution are established to the extent that

- arenas for an interactive dialogue between scientists and policymakers are provided
- deliberations can be transferred to informal arenas without seriously jeopardising the scientific authority of the knowledge base, and
- policymakers acknowledge scientists' authority as a basis for resolving conflicts in questions related to the interpretation and presentation of scientific knowledge.

The question of whether these functions can be combined remains. This is particularly relevant to the maintenance of scientific autonomy/integrity and the level of integration of science and politics. As we have seen, these two functions are assumed to some extent to be linked to the same institutional devices: first and foremost the main function of the scientific bodies. Whereas the maintenance of scientific autonomy/integrity is assumed to depend upon the extent to which scientific bodies primarily do research/research co-ordination, the level of integration of science and politics is assumed to depend upon some combination of research/research co-ordination and formulation of policy implications and advice. This suggests that we to some extent may expect an inverse relationship where one function is served at the expense of the other.

On the other hand, the integration of science and politics is not assumed necessarily to imply integration at all levels. Dimension 9 above, for instance, suggests merely that arenas for an interactive dialogue are provided; they need not be the same as the arenas where the substantive content of the knowledge base is developed. This, therefore, suggests that a separation of science and politics in combination with a certain level of integration of science and politics is at least theoretically possible, as long as the two functions are served by different bodies within the same institutional framework. The establishment of such institutional "buffers" requires a differentiation of institutional arrangements in accordance with the main function of the body. In this respect a balance between scientific autonomy and integrity and a certain level of integration of science and politics may be achieved. This is, however, evidently a difficult task, since what is achieved at one level in terms of maintenance of scientific autonomy/integrity may be counteracted by what is achieved at another level in terms of level of integration of science and politics.

This suggests that while these various functions are linked to specific institutional arrangements, the effect of these arrangements in terms of the capacity of the institutional apparatus to serve these functions may be contextual, depending also on other (situational) factors. This implies, for instance, that the capacity of the institutional apparatus to serve these functions is a matter of degree, and that the extent to which these functions are served may vary strongly even between cases in which institutions have comparable designs. One factor that contributes to this variable effect of institutional design is actor behaviour. Actors are, evidently, not marionettes completely controlled by institutional arrangements, and there is therefore no "one-to-one" relationship between institutional design and actor behaviour. One form of actor behaviour, however, may be particularly important for the manners in which institutional arrangements work and the extent to which the institutional apparatus is capable of serving these functions: namely, performance of leadership in the system.

An institutional framework set up to serve both to separate science and politics and to ensure a certain level of integration of science and politics poses a difficult challenge to the actors operating within the system. The extent to which actors handle these challenges effectively, and hence the extent to which institutional design does contribute to more effective science–policy interactions, may depend upon the extent to which actor behaviour is guided by effective leadership performance. Thus it is not assumed that the capacity of the institutional apparatus to serve these functions solely is dependent upon institutional design. An important additional component is assumed to be actor behaviour taking the form of leadership performance.

4.5 Leadership Performance

The role of individuals and groups acting as leaders has been given attention both in negotiation analysis (Young, 1991; Underdal, 1991, 1994; Malnes, 1995) and in studies of "advisory" science (Mulkay, 1976; Jasanoff, 1990; Cole, 1992). These studies constitute the basis used here for investigating the potential role of leaders in the context of science–policy interaction.

4.5.1 Leadership Performance in Negotiations

In the context of negotiation analysis, leadership has been defined as "an asymmetrical relationship of influence in which one actor guides or directs the behavior of others toward a certain goal over a certain period of time" (Underdal, 1994: 178). Leadership, therefore, is a relationship between leader and followers. Generally, the strength of this relationship may be understood "as a function of the supply of and demand for leadership services" (Underdal, 1994:181). Several modes of leadership, differentiated by the mechanisms by which leadership is provided and the actor capabilities required to succeed, have been identified: Young differentiates between entrepreneurial, intellectual and structural leadership (1991); Underdal differentiates between leadership by unilateral action, coercive and instrumental leadership (1991a,b; 1994); and Malnes distinguishes between problem-solving and directional leadership (1995).

The mechanism by which structural leadership, as described by Young, is provided is mainly "sticks-and-carrots", or threats and promises, and its success depends on the various (structural) power capabilities at the disposal of the agent performing the leadership role and his or her skills in converting this potential into actual influence. Leadership by unilateral action, as described by Underdal, is based on the ability of an actor to "go alone", for instance in the provision of a collective good. The mechanisms by which this mode of leadership is provided are linked to changes in the substantive options available to other actors (in terms of increasing or restricting the scope of possible actions) following from unilateral actions and to the social persuasion unilateral actions may imply. Neither of these modes of leadership are considered relevant in processes of science–policy interaction, since these modes of leadership are not assumed to be in demand in this particular context. Instrumental leadership, on the other hand, may play an important role in a science–policy context. Instrumental leadership which essentially covers two modes of leadership: entrepreneurial leadership (referred to as problem-solving leadership by Malnes, 1995) and intellectual leadership (referred to as directional leadership by Malnes, 1995).

Whereas a structural leadership is based on the leader's ability (power) to force other actors to do something they otherwise would not do, an instrumental leader guides or directs the behaviour of others towards a collective goal mainly by changing the basic parameters of the decision situation (Underdal, 1994; Malnes, 1995). The basic parameters of negotiations fall into two main categories: some that pertain to the interests, values and beliefs of agents and others that are linked to the structure of the interaction (Malnes, 1995). As argued by Malnes, "[a] leader differs from an agent engaged in ordinary bargaining by the fact that he or she does not see these parameters as fixed constraints. The leader tries to alter them so as to bring about an outcome that agrees with collective goals" (Malnes, 1995; 100). Generally, therefore, instrumental leadership is directed towards the substance of negotiations (the interests, preferences, perceptions and beliefs of actors) and/or the institutional setting of negotiations (physical location, rights of participation, rules of proceedings and the method of decision-making).

More specifically, there are two modes of conducting instrumental leadership: Entrepreneurial leadership is directed towards the engineering of politically feasible solutions (Underdal, 1994). The engineering of politically feasible solutions may be undertaken by rearranging the agenda, mainly by linking and de-linking agenda items, so that the joint gain is increased and integrative solutions may be discovered and developed, or by altering the institutional setting to provide incentives for adopting an active problem-solving strategy and to facilitate the integration and aggregation of actor interests (Underdal, 1989; 1990). These two leadership strategies are clearly interrelated and are usually undertaken simultaneously.

In contrast, an intellectual leader influences and guides behaviour more indirectly through "awareness-raising". By providing information, knowledge and guidance more generally, an intellectual leader may influence other actors' beliefs and perceptions of objectives (Young, 1991; Malnes, 1995). Thus, Young identifies an intellectual leader as "an individual who produces intellectual capital or generative systems of thought that shape the perspectives of those who participate in institutional bargaining" (Young, 1991: 298). Malnes defines this mode of leadership more broadly as the ability to "exercise influence" or "to direct other people's behaviour" by "moulding attitudes" and "thereby affect[ing] both the position states take at the negotiation table and their determination to come there at all" (Malnes, 1995: 101). An intellectual leader may guide behaviour through several possible modes of influence, including formal or informal position of authority, cognitive resources (competence, skill and knowledge), or personal qualities (Malnes, 1995: 92). Malnes's definition of this mode of leadership, therefore, is closely linked to an intellectual leader's source of influence, or the capabilities required for leadership efforts to succeed.

Instrumental leadership (both entrepreneurial and intellectual) is based on three capabilities: skills, energy and status (Underdal, 1994). While both substantive and political skills are necessary to perform this mode of leadership successfully, skill alone is not a sufficient condition. The success of the effort also depends on the amount of human energy that is brought to bear on the problem in question, and the formal or informal status, authority or reputation of the individual performing the leadership role or the collective entity the individual represents. Thus, while leadership is exercised by individuals, the capabilities required to succeed are also linked to the formal position or the status of the collective entity (government, agency, organisation etc.) a potential leader represents. Hence, the success of efforts to provide leadership depends on an individual's ability to convert a leadership potential into actual leadership behaviour. Moreover, since informal positions of authority also are considered a potential source of leadership performance, instrumental leadership may, but *need not,* be provided by individuals holding formal positions in the negotiations. This point is of particular importance in this context since, as will be shown below, informal positions of authority are assumed to play a significant role within scientific research communities.

4.5.2 The Role of Scientific Elites

In the scientific community's connections and communications with the wider society, scientific élites have been found to play an important role (Mulkay, 1976). Mulkay argues, first, that the academic scientific research community generates its own well-defined élite: "the members of which are widely regarded by fellow scientists as having made a major contribution to scientific knowledge" (1976: 462). Having been created by social processes internal to the scientific community, the élite remains "strongly committed to objectives regarded as central within that community" (Mulkay, 1976: 446). In the scientific community's communications with the wider society, this élite has been found to serve two main functions: First, much of the pressure exerted by governments and other lay groups on the development of scientific knowledge is mediated through the élite. Second, Mulkay finds that at least in Britain "this scientific élite operates as a 'buffer group', successfully resisting instrumental demands from outside and maintaining considerable freedom for members of the academic research community to pursue their own 'scientifically defined' interests" (Mulkay, 1976: 446).

The scientific élite has also been found to play an important role in the scientific community's production of new knowledge, notably in the selection of new scientific contributions to be regarded by the scientific community as acceptable and important contributions to knowledge:

> Because the criteria of scientific adequacy and value within any research area are so specialized and subtle, the task of deciding which contributions are acceptable and which are important tends to be given to those who are already recognized as having contributed significantly to the field, and, therefore, as having command of the necessary intellectual skills. (Mulkay, 1976: 454)

In this regard the élite seems to play an important role in the development of scientific consensus within a community of scientists. This finding is supported by Cole, who finds that in the evaluation of new scientific work the view of each scientist is influenced by the opinion of others, and that intellectual authorities have more influence than others in this regard: "[intellectual authorities] ...determine which discoveries among a group of contenders for the attention of the scientific community will be successful" (Cole, 1992: 30). This does not imply an unlimited scope for idiosyncratic influence on what the scientific community accepts as adequate and important findings, however, since "authorities who promote work which others see as inferior, ... , can lose their legitimacy" (Cole, 1992: 30).

The scientific élite, therefore, seems to play a role which could be defined as leadership, in accordance with Underdal's definition cited above. More specifically, the leadership of scientific élites seems to be linked to two functions: (i) the development of consensus within the scientific community, where the élite seems to be influential with regard to which new scientific contributions are accepted by the scientific community as "core knowledge" (see also chapter 3, section 3.2.2); and (ii) the scientific community's communications with the wider society, where the élite seems to serve a function as a mediator and as a "buffer group", especially between the scientific community and political authorities and governments.

4.5.3 Leadership Functions in Processes of Science–Policy Interaction

In processes of science–policy interaction there are several tasks for which leadership is likely to be in demand. First, leadership can be assumed to be in demand in the scientific community's evaluation of the adequacy by which state-of-the-art knowledge is represented in the policymaking context. The presentation of scientific knowledge in a policymaking context is always a representation of a (broader) knowledge base external to the particular policymaking process. Given the scientific élite's influential role in the development of core knowledge within scientific communities, its participation in processes of science–policy interaction may be assumed to be important in this regard. We may assume that in a policymaking context scientists are extremely wary of efforts of manipulation and distortion of scientific findings for the sole purpose of supporting policy arguments. Moreover, given the many forms such efforts may take, they may also be difficult to detect, especially for persons not familiar with political tactics. We may thus assume that "bench scientists" look to their élite for guidance in evaluations of the adequacy by which scientific knowledge is represented in a policymaking context. Hence, we may assume that a representation of scientific findings acceptable to the élite also is more acceptable to the scientific community at large than a representation of scientific knowledge opposed by the scientific élite. Moreover, given the élite's scientific merits and reputation for intellectual competence, a knowledge base endorsed by the élite is also likely to be more readily accepted by policymakers.

Second, assuming that scientists in general are unfamiliar with, and perhaps sceptical towards, politics as a system of behaviour, the scientific élite's competence in this regard, given its role as mediator and buffer in the scientific community's regular dealings with political and governmental authorities, may be assumed to be in demand in processes of science–policy interaction. This mode of leadership can take a general form, simply as a guide to behaviour within a complex institutional structure. A more specific mode of leadership can also be envisaged, where scientific élites may serve as a "communicative link" between the scientific and the political bodies of the interaction, instrumental for communicating to policymakers what scientists can and cannot do in terms of the provision of scientific knowledge, and for communicating the needs of policymakers in terms of scientific knowledge to (the community of) scientists. The provision of this mode of leadership may be important to enhance the capacity of the institutional apparatus to combine the separation of science and politics with a certain level of science–policy integration. The scientific élite seems to be particularly qualified and competent for performing this kind of leadership. This assumption is also supported by Jasanoff's findings:

> Given the importance of negotiation and consensus-building in the advisory process, it is clear that the ideal adviser needs to be more than a mere technical expert. ... Agency advisers, ... , must be capable of doing convincing boundary work while they are engaged in making science policy. Such balanced performance calls for experts who not only display an informed sensitivity to the agency's mission but also enjoy unquestioned standing among their peers. (Jasanoff, 1990: 243, see also chapter 3, section 3.3.4.)

Third, the competence of scientific élites to provide leadership in the communication between the scientific and political bodies of the interaction may also be linked to the more vaguely defined need of mechanisms for conflict resolution. While formally institutionalised mechanisms for conflict resolution adopted from regular negotiations, may, as discussed above, be inappropriate within a context of science–policy interaction, the informal services that may be provided by individuals in this regard, and especially individuals that are highly regarded within their various disciplines, may serve the same purpose in a more (scientifically and politically) acceptable way. Situations of conflict or strong disagreement, especially between scientists and policymakers, are precisely the situations in which the provision of leadership based on informal positions of authority (perhaps, but not necessarily, in combination with formal positions of authority) are in demand. The extent to which this mode of leadership is in supply will depend upon the extent to which scientific élites take part in processes of science–policy interaction, and, of course, their willingness and personal capabilities to perform a leadership role of this kind. Their willingness to take on this kind of leadership role may be linked to the extent to which they perceive such a role as a threat to their informal position as a scientific authority. Their personal capabilities to perform this role may be linked to their ability to exert the touch of diplomacy this role may be assumed to require in order to succeed.

Finally, a leadership role, not directly based on the informal authority of the scientific élite, but rather linked to the ability of individuals to interpret and translate scientific findings so that they become more accessible and applicable to policymakers is also likely to be in demand in processes of science–policy interaction. In her studies of the development of an international regime to combat ozone depletion, Karen Litfin finds that "knowledge brokers" – i.e. "intermediaries between the original researchers, or the producers of knowledge, and policymakers who consume that knowledge but lack the time and training necessary to absorb the original research" – play an important role in framing and interpreting knowledge in policymaking contexts, thus making the knowledge more available and applicable to policymakers (Litfin, 1994: 4). Knowledge brokers are typically experts, not scientists active in research, who may work for government agencies, industry, international organisations or non-governmental organisations. Their influence in policymaking is their ability to "set the terms of discourse" (Litfin, 1994: 188). In the ozone process, knowledge brokers were instrumental in the transformation of scientific knowledge and especially scientific methods into tools for decision-making, notably the "chlorine-loading" methodology.[2] Moreover, Litfin emphasises that the leadership functions served by knowledge brokers may be in particular demand – or conversely, knowledge brokers may be particularly influential – in situations characterised by scientific uncertainty where there is no or only a loosely defined scientific consensus within the scientific community, and where policymakers may thus have to orient themselves in a

[2] The "chlorine-loading" methodology is essentially a method of assessing the levels at which chlorine concentrations in the atmosphere damage the ozone layer and, accordingly, the levels of chlorine concentrations at which the ozone layer can recover. By focusing on chlorine concentrations rather than the more complex calculation of ozone depletion potentials (ODPs), the scenario-modelling field was opened up "to anyone with a powerful desktop computer" (Parson, 1993: 63).

landscape of diverging and perhaps conflictual scientific "advice" (Litfin, 1994). This mode of leadership is not necessarily based on informal positions of authority, but rather an individual's position as intermediary between science and politics (for instance a governmental or non-governmental expert), his or her personal qualifications, competence and knowledge within a particular problem area and, above all, his or her ability to take advantage of "windows of opportunity" as and when they arise. In the ozone process the "window of opportunity" that came to have significant implications both for knowledge brokering and the final outcome was the discovery of the Antarctic ozone hole and the fact that this discovery was as much a surprise to the scientific community as it was to policymakers (Litfin, 1994).

4.5.4 Links Between Institutional Design and Leadership Performance

Institutional design and performances of leadership are assumed to be related: first, in the sense that institutions may to a varying extent provide incentives to take on a leadership; and second, institutions may to a varying extent provide potential leaders with the sufficient formal capacity to perform the role effectively. In the case of processes of science–policy interaction, this aspect is linked to the extent to which members of the scientific élite are recruited into formal positions within the institutional structure, which in some cases may be very important for the capability of the scientific élite to succeed in efforts to perform a leadership role. The provision of incentives and capabilities to take on and serve a role as leader successfully are, therefore, also linked to the extent to which the institutional apparatus provides sufficient "room" for potential leaders to provide actual leadership services.

Leadership performance is linked to institutional design also in the sense that the leadership may be directed towards institutional design itself. That is, one leadership function, as discussed above, may be to alter institutional arrangements to enhance the capacity of the institutional apparatus to serve the four main functions.

4.6 Problem Malignancy and the State of Knowledge

It is not assumed that the organisation of the science–policy dialogue and leadership performance in the science–policy interaction process are the only factors that determine the policy response to a particular problem. On the contrary, the process whereby a problem is diagnosed can never determine the outcome of a process which is essentially political in nature. This implies that even under the most favourable conditions for the transformation of scientific knowledge into premises for policy choice, scientific knowledge may still not be acted upon in the policy decisions made. Thus, the extent to which scientific knowledge is acted upon by policymakers is also assumed to be controlled by a set of factors other than the effectiveness of the science–policy dialogue, that are not discussed in this study.

Furthermore, the outcome of the science–policy dialogue may also be a result of other factors than the institutional design of the process and the amount of leadership provided. In the empirical analysis of this study, therefore, the robustness of the relationship between institutional design, leadership performance and the effectiveness of the process indicated by the empirical analysis is controlled against the impact of two main alternative explanations: the extent to which the outcome is produced by the nature of the political setting of the problem, conceptualised in terms of problem malignancy, and the extent to which the outcome can be seen as a result of the state of knowledge in the issue area, conceptualised in terms of its conclusiveness.

4.6.1 Problem Malignancy

This analysis adopts the conceptualisation of problem malignancy developed by Underdal[3] (in Miles et al., subm. [4]), where the political malignancy of a problem is conceived of "primarily as a function of the configuration of actor interests and preferences which it generates" (p. 15). According to this conceptualisation, a perfectly benign problem would be characterised by identical interests. The further away from this state of harmony, the more malign the problem becomes.

International regimes are established to deal with two main categories of problems: problems of incongruity and problems of co-ordination (ibid.). The former may be considered the more malign of the two, and we concentrate our discussion on this category.

The defining characteristic of problems of incongruity is that, "the cost-benefit calculus of an individual actor is systematically 'biased' in favor of either the costs or the benefits of a particular course of action" (p. 17). Such a bias may be due to the (objective) distribution of material consequences, the perspective actors apply when assigning a value to these consequences, or both. Trans-frontier pollution and common pool resources are typical examples of problems of this kind. For instance, in a case where pollution generated in one country causes damage in other countries, the polluting

[3] For another conceptualisation of problem malignancy, see Efinger and Zürn, 1990.

[4] Page references refer to the draft, and may, therefore, not correspond with page numbers as they appear in the publication.

country may harvest all the benefits associated with the (polluting) activity, without bearing any of the costs. Similarly, if the fishing fleet of one country depletes a common pool resource, it would share the costs associated with the depletion (the loss of future fishing opportunities) with all others, while reaping the short-term benefits alone (both examples are from Underdal, ibid.).

The extent to which such asymmetries in the distribution of material consequences become problems of incongruity also depends on the perspectives within which actors assign value to perceived consequences: An altruist, who would be equally concerned with the welfare of others as with that of himself, would incorporate the consequences to others in his cost-benefit calculus. A competitor, by contrast, would be concerned with relative gains and hence assign zero value to the welfare of others.

Problems of incongruity arise when actors under-represent actual costs or actual benefits. This under-representation is linked to at least two different mechanisms: externalities and competition. Externalities refer to "those effects of an actor's behavior that 'hit' others and (therefore) disappear from the actor's own cost-benefit calculations" (p. 19). Competition exists "whenever one actor's (subjective) welfare depends on how well he performs compared to others" (ibid.). Competition is the more malign of the two since "competition can distort actor incentives by *amplifying* the costs of cooperative behavior or by positively *rewarding* defection in ways which pure leaks can never do" (ibid.).

To the extent that problems of incongruity also are characterised by asymmetry and/or cumulative cleavages their malignancy is enhanced. A problem is asymmetrical to the extent that "the parties involved are (or perceive themselves as being) affected differently so that their interests are negatively correlated" (p. 20.). Cleavages are cumulative to the extent that "parties find themselves in the same situation on all dimensions or issues, so that who stand to 'win' (or 'lose') on one dimension also come out as 'winners' ('losers') on the other dimensions as well" (ibid.).

We now have a definition of problem malignancy against which the nature of the climate change problem may be compared in our empirical analysis. A problem is considered malign to the extent that it embodies features of incongruity, where actors under-represent actual costs or actual benefits in their cost-benefit calculus. The level of malignancy is determined by the extent to which this under-representation is caused by externalities or competition – of which the latter is considered the more malign – and the extent to which the problem is characterised by asymmetries and cumulative cleavages.

4.6.2 State of Knowledge

While scientific consensus may be seen as encompassing scientific uncertainty in the sense that scientists agree on what they are uncertain about and hence where they may be in possible disagreement, scientific consensus is more often associated with relatively conclusive knowledge. In this analysis, therefore, state of knowledge is conceptualised in terms of its conclusiveness. Scientific consensus (conclusive knowledge) is thus seen as inversely related to scientific uncertainty. In this regard, the level of scientific uncertainty is considered an indicator of the conclusiveness of the state-of-the-art knowledge represented in the knowledge base provided.

4.7 In Sum

The institutional framework of processes of science–policy interaction has been found to hold an instrumental potential to the extent that,

- institutions for science–policy interaction are formally institutionalised
- they are compatible with the overriding goal of maintenance of the scientific autonomy/integrity of participating scientists, and
- within the limits caused by political conflict.

This instrumental potential is assumed to be realised to the extent that the institutional framework is capable of serving four main functions: maintaining scientific autonomy/integrity, providing a sufficient level of integration of science and politics, ensuring the geographic representativeness of scientific bodies, and providing mechanisms for dealing with conflicts or strong points of disagreement especially between scientists and policymakers. These functions, moreover, are assumed to be linked to a set of more specific institutional arrangements including recruitment and funding mechanisms, the operational autonomy of scientific bodies, the main function of scientific bodies, and the extent to which arenas for an interactive dialogue between scientists and policymakers are provided. Finally, it was argued that the capacity of the institutional apparatus to serve these functions is not solely dependent upon institutional design. An important additional factor is assumed to be the extent to which actor behaviour taking the form of leadership performance occurs. The following three categories of leadership functions are assumed to be in particular demand in processes of science–policy interaction:

1. Leadership functions directed towards the development of the knowledge base – particularly with regard to the scientific community's evaluation of the adequacy by which state-of-the-art knowledge is represented in the knowledge base.

2. Leadership functions directed towards the transformation of the knowledge base into premises for decision-making.
3. Boundary role leadership directed towards the provision of communicative links between scientists and policymakers as well as the development of institutional arrangements.

Leadership directed towards the provision of informal mechanisms for conflict resolution is assumed to constitute an integral part of all of these modes of leadership, but particularly categories 1 and 2.

The relationship between the dependent and independent variables is "tested" against the likely impact of two main factors: the political malignancy of the problem and the state of knowledge. The former factor is conceptualised in terms of the extent to which the problem has the features of a problem of incongruity. The latter factor is conceptualised in terms of its conclusiveness.

Chapter 5

The Development of an International Regime on a Human-Induced Climate Change

5.1 Introduction

In this chapter, the problem of a human-induced climate change and the scientific and historical background for the development of an international regime addressing this problem is presented (sections 5.2 and 5.3). In section 5.4 the effectiveness of the associated science–policy dialogue is assessed according to the conceptualisation of effectiveness developed in chapter 2. The Intergovernmental Panel on Climate Change (IPCC) is organised so that all IPCC reports must be accepted and/or approved by both scientists and policymakers. Thus, in this assessment effectiveness is judged in relation to what is agreed upon within the IPCC as well as in relation to what is agreed upon in the UN Framework Convention on Climate Change.

5.2 The Problem of a Human-Induced Climate Change[1]

The ultimate energy source for weather and climate on Earth comes from the sun. The Earth intercepts solar radiation. Approximately 30% of solar radiation is reflected by air, clouds and the Earth's surface. Approximately 25% is absorbed by the atmosphere. The rest is absorbed by the Earth's surface and warms it. The energy absorbed from solar radiation is balanced (in the long term) by outgoing long-wave infrared radiation. Infrared radiation is absorbed in the atmosphere by clouds and greenhouse gases. Thus, some of the long-wave terrestrial radiation emitted by the warm

[1] This section is based on IPCC, 1990; 1994; 1995; and Torvanger et. al., 1997.

surface of the Earth is partially absorbed by the atmosphere and then re-emitted by a number of trace gases in the cooler atmosphere above. The absorption of infrared radiation in the atmosphere is primarily caused by water vapour (H_2O), carbon dioxide (CO_2) and clouds, but methane (CH_4), nitrous oxide (N_2O) and ozone (O_3) also contribute significantly to this effect. This is the natural greenhouse effect, which keeps the Earth's surface temperature approximately 34°C warmer than it would be had these natural greenhouse constituents not been present in the atmosphere. Figure 5.1 shows a simplified illustration of the greenhouse effect.

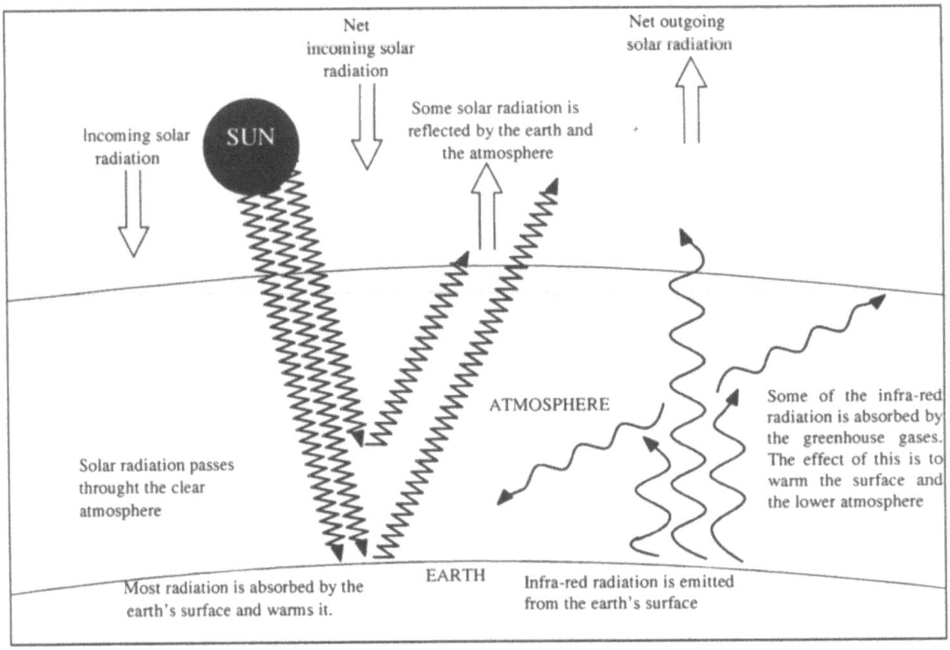

Figure 1. The greenhouse effect. Source: IPCC, 1994:15

Since the pre-industrial era the atmospheric concentration of carbon dioxide has increased by approximately 30%. The concentration of methane has increased by 145%, and the concentration of nitrous oxide has increased by approximately 15% over the same period. There is broad agreement that this increase in concentrations is largely a result of human activities. Human activities have also added new greenhouse gases that are not naturally present in the atmosphere (for instance chlorofluorocarbons – CFCs).

The human-induced increase in atmospheric concentrations of greenhouse gases has caused an *enhanced* greenhouse effect by changing the radiative balance of the Earth–atmosphere system. A change in the radiative balance is often referred to as radiative forcing (usually measured in Wm^{-2}). A positive radiative forcing tends on average to warm the surface, while a negative radiative forcing tends on average to cool the surface. The enhanced greenhouse effect is expected to cause a rise in the global-average, annual-mean surface-air temperature (for simplicity referred to as the global mean temperature) and changes in other climate parameters (such as precipitation, humidity, sea level, winds, etc.). The rate of change depends, *inter alia*, on inertia in the system, the rate in which emissions change, feedback effects, etc., and is, therefore, uncertain.

Increased atmospheric concentrations of CO_2 are most important for the human-induced (enhanced) greenhouse effect (approx. 60%). Anthropogenic emissions of CO_2 are primarily caused by the combustion of fossil fuels and deforestation in tropical areas. Halocarbons, CH_4, tropospheric ozone and N_2O also contribute substantially (see Fig. 5.2.).

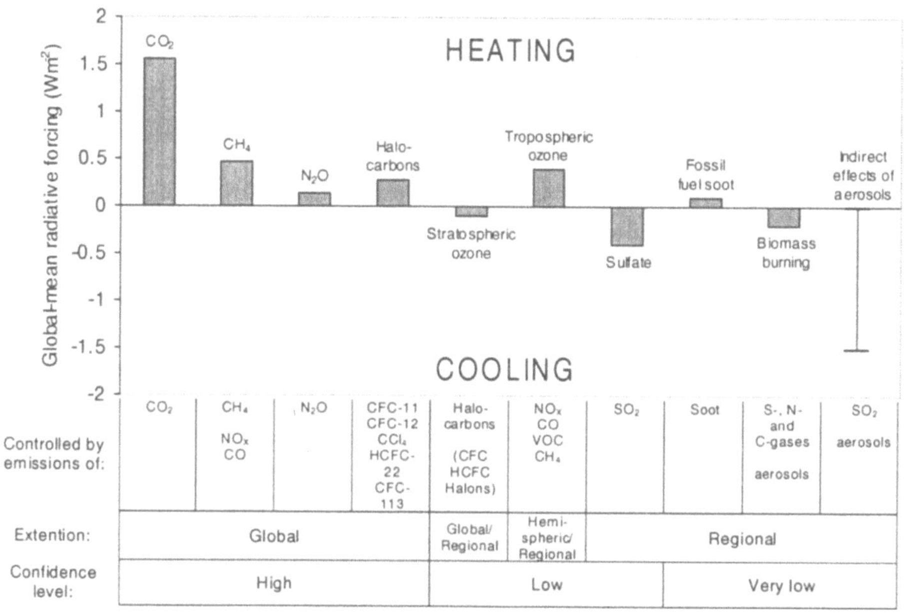

Figure 2. Source: Skodvin and Fuglestvedt, 1997.

Emissions of sulphur dioxide (SO_2) are converted in the atmosphere into aerosols, which reflect solar radiation back to space and thereby reduce incoming solar radiation. Emissions of SO_2 thus cause a negative radiative forcing which is assumed to have modified the warming effect caused by increased concentrations of greenhouse gases. The cooling effect caused by SO_2 is, however, regional in nature due to the short atmospheric lifetimes of SO_2 and particles. While increases in ozone in the troposphere (the lowest 10–15 km of the atmosphere) have had a warming effect, reductions in ozone in the stratosphere (the "ozone layer") have probably had a cooling effect.

Since around 1800, the global mean temperature has increased by 0.3–0.6°C. Available evidence suggests that the 20th century global mean temperature is at least as warm as any other century since at least 1400 AD. Several studies show that the observed warming trend is unlikely to be entirely natural in origin. Scientists are still cautious about attributing this warming trend to human activities since there is significant uncertainty concerning natural climate variability. In 1995, however, the IPCC concluded that "the balance of evidence suggests that there is a discernible human influence on global climate" (Summary for Policymakers, 1995:11).

Climate models project an increase in global mean temperature of 1–3.5°C by 2100. Because of the thermal inertia of the oceans, only 50–90% of the eventual equilibrium temperature change will have been realised by 2100, and temperature will continue to increase beyond 2100, even if concentrations of greenhouse gases are stabilised by that time. The projected increase in global mean temperature is assumed to be associated with a 15–95 cm sea level rise. Moreover, sea level is expected to continue to rise at a similar rate in future centuries beyond 2100, even if atmospheric GHG concentrations are stabilised by that time, and even after the global mean temperature has stabilised. Ocean circulation patterns could also change.

A system's sensitivity to changes in climate and its ability to adapt to such changes determines the system's vulnerability to changes in climate. A rise in global mean temperature could have potentially significant adverse effects on terrestrial and aquatic ecosystems, the global hydrological cycle and regional water resources, food production, human infrastructure and human health. The reliability of regional-scale predictions of impacts is low. Generally, however, vulnerability is assumed to depend on economic circumstances and institutional infrastructure, which would imply that systems are typically more vulnerable to changes in climate in developing countries where the economic and institutional circumstances are less favourable.

It should be emphasised that the problem of a human-induced climate change is characterised by significant scientific uncertainty with respect to demonstrating that an observed climate change is statistically unusual (detection), establishing cause-and-effect relations (attribution) and, not least, projecting future change. Significant uncertainty is linked, among other things, to:

- estimates of future emissions and biogeochemical cycling (including sources and sinks) of greenhouse gases, aerosols and aerosol precursors;
- feedbacks associated with clouds, oceans, sea ice and vegetation and hence representation of climate processes in models; and
- collection of long-term observations of solar output, atmospheric energy balance components, hydrological cycles, ocean characteristics and ecosystem changes for the purpose of model testing, assessment of temporal and regional variability and for detection and attribution studies.

5.3 The Scientific and Political History of the Development of an International Regime on a Human-Induced Climate Change

Scientific recognition of the potential of human activity to modify climate dates back to the early 19th century. In 1827, Baron Jean-Baptiste Fourier suggested that human activity can modify surface climate, and was perhaps one of the first to suggest the now well-known greenhouse effect of the atmosphere (Ramanathan, 1988). The greenhouse theory of climate change was taken up in earnest by the Swedish scientist Svante Arrhenius. His first estimate of a man-made global temperature change caused by industrial emissions was published in 1896 (Rodhe et. al., 1997). Scientific interest in man's potential impact on global climate was activated in the 1950s when monitoring of atmospheric concentrations of carbon dioxide began in Antarctica and Hawaii (Agrawala, 1998a). This interest was further mobilised through conferences, loose research networks and assessments especially from the 1970s onwards (Agrawala, 1998a). There have been brief periods when scientific interest has focused on the possibility of a global cooling trend (due to the cooling effect of industrial aerosols), but by the early 1980s the predominant interest of the scientific community converged on the possibility of warming trends in global mean temperature caused by increased atmospheric concentrations of greenhouse gases (Agrawala, 1998a).

The starting point for international efforts to better understand climate variations and the possible problem of a human-induced climate change is generally regarded to be the UN Conference on Human Development in Stockholm in 1972. In 1979 the World Climate Conference was held in Geneva, and the World Climate Programme (WCP) was launched. The creation of the WCP set forth a series of workshops organised under the auspices of the World Meteorological Organization (WMO), the United Nations Environment Programme (UNEP) and the International Council of

Scientific Unions (ICSU) held in Villach, Austria, in 1980, 1983 and 1985 (Agrawala, 1998a). At the 1985 Villach meeting an international group of scientists reached a consensus that, as a result of the increasing concentrations of greenhouse gases in the atmosphere, a rise in the global mean temperature "greater than any in man's history" could occur in the first half of the next century (WMO, 1985: 1). This group of experts also stated that "the understanding of the greenhouse question is sufficiently developed that scientists and policymakers should begin active collaboration to explore the effectiveness of alternative policies and adjustments" (WMO, 1985: 3).

In combination with a set of other factors, especially anomalous weather conditions in Europe and America, the 1985 Villach meeting was instrumental in bringing the climate issue onto the international political agenda. In 1986 the Advisory Group on Greenhouse Gases (AGGG) was set up under the joint sponsorship of WMO, UNEP and ICSU. Each of these bodies nominated two experts. The panel consisted of seven members: Gordon Goodman, Bert Bolin, Ken Hare, G. White, G. Golitsyn, Sukiyoro Manabe and M. Kassas (Agrawala, 1998a).

During the latter half of the 1980s the climate issue increasingly gained saliency among the public, scientists and policymakers. At the Toronto Conference of the Atmosphere, where more than 300 scientists and policymakers from 48 countries, UN organisations, IGOs and NGOs participated, an explicit policy recommendation calling upon national governments to reduce their CO_2 emissions by 20% from 1988 levels by 2005 was agreed upon.

Meanwhile, the WMO and UNEP in close co-operation with various US agencies agreed that an *intergovernmental* mechanism was needed to undertake further internationally co-ordinated scientific assessments of climate change, and invitations to governments to the first session of the IPCC were sent out early in 1988 (personal communication with Bert Bolin; for a detailed account of the history behind the establishment of the IPCC see Agrawala, 1998). The AGGG set up in 1986 was gradually replaced by the IPCC and has not met since 1990.

The first plenary session of the IPCC took place in November 1988. The objective of the IPCC, formulated by the governing bodies of WMO and UNEP, is twofold: (i) To assess the scientific information related to the various components of the climate change issue and the information needed to evaluate the environmental and socio-economic consequences of climate change; and (ii) to formulate "realistic response strategies for the management of the climate change issue". In 1988, three Working Groups (WGs) were set up to attain this objective: Working Group I to assess available scientific information on climate change, Working Group II to assess environmental and socio-economic impacts of climate change, and Working Group III to formulate response strategies (Report of the first session of the IPCC). In 1992, the IPCC was reorganised: WGs II and III were merged in the old WGII, while a new WGIII was set up to deal with socio-economic and other cross-cutting issues related to climate change. In 1990 the First IPCC Assessment Report was presented to the Second World Climate Conference, where it was accepted as an adequate basis on which to start negotiations. The IPCC plenary accepted the Second IPCC Assessment Report in December 1995. In between the first and the second assessment reports, the IPCC has provided two interim reports in 1992 and 1994. Work on a third assessment is underway (current work plans suggest finalisation in 2001).

In December 1990 the Intergovernmental Negotiating Committee for a Framework Convention on Climate Change (INC/FCCC) was formally established in UN General Assembly Resolution 45/212. Thus, the political process is under the direct control of the General Assembly while the IPCC is organised under the auspices of WMO and UNEP. In February 1991 the first session of the INC was convened. From February 1991 to April 1992, five sessions were held. The fifth session resulted in a Climate Convention ready for signing at the United Nations Conference on Environment and Development (UNCED) in Rio, June 1992. The United Nations Framework Convention on Climate Change (UNFCCC) entered into force in March 1994.

The stated ultimate objective of the 1992 UN Framework Convention on Climate Change is "to achieve...stabilization of greenhouse gas concentrations in the atmosphere at a level that would prevent dangerous anthropogenic interference with the climate system" (Art. 2). Each of the Parties to the Convention have committed themselves to, "adopt national policies and take corresponding measures on the mitigation of climate change, by limiting its anthropogenic emissions of greenhouse gases and protecting and enhancing its greenhouse gas sinks and reservoirs" (Art. 4.2 (a)). Moreover, Parties are committed to communicate to the Conference of the Parties (COP) "detailed information" on such policies and measures and the resulting projected emissions of greenhouse gases "with the aim of returning individually or jointly to their 1990 levels these anthropogenic emissions of carbon dioxide and other greenhouse gases not controlled by the Montreal Protocol" (Art. 4.2 (b)). Finally, the Parties agreed to review the adequacy of these commitments at the first session of the COP (Art. 4.2 (c)). The outcome of the first session of the COP, held in Berlin in March/April 1995, was the Berlin Mandate, in which the parties acknowledged the inadequacy of the commitments adopted in the Climate Convention and agreed to begin the process of strengthening them in an additional protocol or another legal instrument to be signed at the third session of the COP in December 1997. The Ad hoc Group on the Berlin Mandate (AGBM) was set up to address this task.

The work of the AGBM resulted in the Kyoto Protocol to the climate convention, agreed upon at the third session of the COP in December 1997. The Kyoto agreement implies a legally binding commitment by industrialised countries to a total GHG emissions reduction of 5.2% by 2012 as compared to emission levels in 1990.

5.4 The Effectiveness of the Science–Policy Dialogue

What level of effectiveness of the science–policy dialogue does this outcome suggest? To recall, the dependent variable is defined in terms of three levels (chapter 2), two of which are related to the extent to which policymakers acknowledge the scientific authority of the knowledge base and accept its substantive conclusions as factually valid, and one that is related to the extent to which policymakers act upon this knowledge base in the decisions made. Policymakers' acceptance is assumed to be reflected in the development of a consensual problem diagnosis. At the first level, a knowledge base, representative of state-of-the-art knowledge within relevant fields and disciplines, which describes the basic cause-effect relationships of the problem is developed. At the second level, scientists and policymakers also agree on how this knowledge base is linked to valued policy goals. At the third level, policymakers adopt this problem diagnosis as a premise for the decisions made and act upon (the policy implications of) its substantive conclusions.

The effectiveness of the science–policy dialogue in the case of climate change clearly does not correspond to what we have defined as most effective (level three). Seven years after the adoption of the first IPCC Scientific Assessment Report (1990), governments have committed themselves only to very limited reduction targets for emissions of GHGs.[2] The scientific knowledge base provided by the IPCC, therefore, has only to a very limited extent been followed by a policy response in terms of actions by governments to restrict the accumulation of GHGs in the atmosphere. The commitments to reduce emissions of GHGs agreed upon in the Kyoto Protocol will only have a marginal to negligible effect in this regard (Bolin, 1998). Therefore, this agreement does not qualify as an adoption of the scientific knowledge base in the sense that policymakers have fully recognised the policy implications of the problem of a human-induced climate change and acted accordingly.

[2] It should be noted that this time span is not unusual, however. In comparison, it took 13 years before the first protocol with legally binding commitments to reduce SO_2 emissions was agreed upon from the time when the acid rain issue first surfaced on the international political agenda (1972–1985).

When we turn to policymakers' acceptance of the knowledge base provided by the IPCC, the situation is different (level one and two in the definition of effectiveness). In general, the signing (1992) and entering into force (1994) of the UNFCCC is an indication that the problem of a human-induced climate change is accepted by governments as constituting a political problem whose resolution has to be found in co-ordinated action at the international level. Article 2 of the Convention states its objective to be a stabilisation of GHG concentrations in the atmosphere that prevent "dangerous" anthropogenic interference with the climate system, and, moreover, that this stabilisation should be achieved within a time-frame allowing ecosystems to adapt naturally to climate change. It is important, however, to note the UNFCCC usage of the term "climate change" as "...a change of climate which is *attributed directly or indirectly to human activity* that alters the composition of the global atmosphere and which is in addition to natural climate variability observed over comparable time periods" (Art. 1.2, emphasis added). The IPCC usage of this term *does not* imply attribution to human activities. Hence, the objective of the UNFCCC hinges on the attribution of observed climate change to human activities. In 1995 (three years after the signing and one year after the entering into force of the Convention), the IPCC, cautiously, attributed the observed climate change to human activities in reporting that "the balance of evidence *suggests* a discernible human impact on climate" (SPM, 1995: 11, emphasis added). While the UNFCCC definition of climate change may indicate that the "burden of proof" rests on scientists (to establish a "strong" causal link between human activities and observed climate change), a precautionary principle has, however, also been adopted as a guiding principle (Art. 3).

Turning more explicitly to the indicators of effectiveness suggested in chapter 2, however, we may conclude that the knowledge base on a human-induced climate change has been accepted by policymakers in the sense that all IPCC reports have been accepted and/or adopted by the full IPCC plenary, in which governments are represented. Thus, as a function of the institutional structure of the IPCC, IPCC reports represent the "consensus" view of both scientists and policymakers. Having succeeded in providing these reports, therefore, the IPCC effort, and hence the science–policy dialogue of this case, has been effective in terms of our level one definition of effectiveness.

The question of whether the IPCC effort also has succeeded in terms of linking this knowledge base to valued policy goals in a manner agreed upon by both scientists and policymakers is more complex. The objective of the Convention (Art. 2) may be regarded as an indication that an atmosphere with GHG concentrations that do not exceed "dangerous" levels (in terms of anthropogenic interference with the climate system) is a valued policy goal.

Moreover, the increased atmospheric concentrations of the GHG that contributes most to the enhanced greenhouse effect, CO_2, is linked to human activities, namely the combustion of fossil fuels and land-use changes causing deforestation. This suggests that reductions in CO_2 emissions from human activities (primarily the combustion of fossil fuels) have to be part of a solution to the problem.[3]

In article 1.3 of the Convention, however, it is stated that policies and measures to deal with climate change should be "comprehensive". This implies that stabilisation of atmospheric GHG concentrations below a "dangerous" level can be achieved also by reducing emissions of other GHGs than CO_2. The source of some of these "other" GHGs, in terms of societal activities, is uncertain or unknown.[4] A comprehensive approach also requires methods by which the warming effect of different GHGs can be compared. The current method for comparing different GHGs is the calculation of Global Warming Potentials (GWPs[5]). In the application of GWPs (for policymaking), however, the highly varying atmospheric lifetimes of GHGs is a difficult issue, with potentially significant policy implications. In the Kyoto Protocol, however, a 100-year time horizon is adopted for calculations of GWPs. With a fixed time scale, the calculation of GWPs may serve as a policy instrument for implementing a comprehensive approach (for a discussion of the GWP methodology see Skodvin & Fuglestvedt, 1997; Shackley & Wynne, 1997; Smith and Wigley, in press; Fuglestvedt et al., subm.).

The UNFCCC also ascribes the main responsibility for the adoption and implementation of commitments mainly to developed countries (OECD members). This is specified in Article 3.2, and in the distinction made between the commitments of Annex I parties (developed countries and countries with economies in transition), Annex II parties (developed countries, i.e., OECD members) and developing country parties. This

[3] In IPCC, 1990, this point was *illustrated* in the statement "calculated with confidence" that "the long-lived gases would require immediate reductions in emissions from human activities of over 60% to stabilise their concentrations at today's levels" (Executive Summary, IPCC 1990). This is usually interpreted as referring to CO_2. Reforestation has also been discussed as a possible means by which to cope with this problem. The effect of reforestation is, however, difficult to quantify and is also problematic in terms of baselines and time horizon.

[4] While the anthropogenic sources of methane emissions are well known, their effect in terms of increased atmospheric concentrations is difficult to quantify. Anthropogenic sources of nitrous oxide emissions are less known and hence their effect is even more difficult to quantify.

[5] Calculations of GWPs for some species are problematic and controversial, notably gases whose lifetime is shorter than the time for mixing in the troposphere. The method has, however, attained acceptance as a tool for policymaking within the political and scientific communities (see Skodvin and Fuglestvedt, 1997).

principle was contested by the USA in the negotiation process towards the third COP meeting where the Kyoto Protocol to the UNFCCC was adopted. The principle is, however, retained also in the Kyoto Protocol.

Finally, options for reducing emissions of GHGs have been assessed by the IPCC (WGII). The Second Assessment of the IPCC (1995) includes the assessment of options linked to:

- energy use, improved energy efficiency;
- energy supply, GHG reductions in use of fossil fuels – for instance through efficiency improvements in conversion of fossil fuels, a switch to low-carbon fossil fuels (such as oil to gas), as well as switching to non-fossil fuel sources of energy; and
- management of forests, agricultural lands and rangelands, including sustaining existing forest cover and slowing down deforestation.

Possible pathways for reducing GHG emissions are therefore suggested and agreed upon by scientists and policymakers in IPCC reports, although these options are listed without priority. There is also no indication whatsoever of the possible pathways for which there might be political agreement. Most likely, however, governments will employ a mix of these options which will vary between countries.

5.5 In Sum

This discussion indicates that the science–policy dialogue on climate change has been fairly effective. A problem diagnosis seems to have been developed in terms of both of its constitutive elements: A knowledge base, accepted by both scientists and policymakers as an adequate representation of state-of-the-art-knowledge in relevant fields has been developed, and this knowledge has been linked to valued policy goals. This linkage is established by the attribution of (problem) cause to societal activities and the suggestion of alternative pathways to possible solutions. This high level of acceptance is also reflected in a moderate behaviour change. In the Kyoto Protocol of 1997, policymakers have made some (rather weak) obligations to reduce their GHG emissions. Although based on science, these commitments can not be considered to be an adoption of research-based knowledge, however. Rather, these commitments are interpreted as a confirmation of policymakers' acceptance of the factual validity of the knowledge base provided by the IPCC. It thus seems reasonable to conclude that the science–policy dialogue seems to have been effective in terms of a generally accepted and recognised problem diagnosis, while it has to a lesser extent been effective in terms of explicit actions and a behaviour change following from this recognition of the problem.

Chapter 6

Structure: The Institutional Design of the Intergovernmental Panel on Climate Change

6.1 Introduction

This chapter explores the extent to which the institutional framework within which the science–policy dialogue on climate change takes place has contributed to this outcome. To what extent can the level of effectiveness described above be explained by way in which the science–policy dialogue is organised?

The analysis in chapter 4 suggests that institutional design could enhance the effectiveness of the science–policy dialogue to the extent that the institutional apparatus is designed to serve four main functions: maintain the scientific autonomy and integrity of the scientific bodies; ensure a certain level of involvement and interaction between scientists and policymakers; ensure a geographic balance in participation; and provide mechanisms for conflict resolution.

The main question addressed in this chapter, then, is twofold: To what extent is the institutional apparatus of the science–policy dialogue on climate change taking place within the framework of the Intergovernmental Panel on Climate Change capable of serving these functions? And what kind of impact does this have on the effectiveness of the science–policy dialogue? This question is approached in three steps: First, a general overview of the institutional set-up of the IPCC and a more detailed account of the assessment process is provided in sections 6.2 and 6.3, with the main focus on the proceedings of WGI contrasted by a brief discussion of the proceedings of the old WGIII. In section 6.4 the capacity of the institutional apparatus of the IPCC to serve the four main functions is discussed and analysed according to the indicators suggested in chapter 4. Finally, in section 6.5, a judgement is made about the extent to which the institutional apparatus of the IPCC has served these functions, and the extent to which the IPCC's score in this regard may explain the outcome as described above.

6.2 The Institutional Set-up of the IPCC

The IPCC constitutes the scientific body of the climate change regime, while the Conference of the Parties to the Climate Convention (COP) constitutes the political or regulatory body. However, the Executive Council of WMO and the Governing Council of UNEP established the IPCC as an intergovernmental organisation under UN auspices. Being a UN body, the IPCC is submitted to the traditional UN procedures governing most UN bodies. Thus, one main characteristic feature of the IPCC is that it has a scientific mandate, while it is organised within a political institutional framework. This is reflected in a distinction within the IPCC itself between bodies largely serving administrative/political functions and bodies constituting the scientific/technical core (Figure 7.1). The IPCC thus operates, in the most literal sense of the word, in the interface of science and politics. The distinction between science and politics within the IPCC constitutes a zone rather than a clear-cut border.

The main decision-making bodies of the IPCC are the plenary, the bureau, three working groups (WGs) (some with ad hoc task force establishments on specific issues) and, from 1989 to 1992, a Special Committee for the participation of developing countries. The scientific/administrative distinction runs through the WG level, between WG plenary (as distinguished from panel plenary) and task force establishments and contributing - and lead-author meetings, which constitute the scientific core.

The main function of the IPCC is to provide scientific Assessment Reports of state-of-the-art knowledge on the various aspects of a possible human-induced climate change. WGI is responsible for providing assessments on the science of climate change, while WGII provides assessments on the impacts of climate change. Before 1992, WGIII was responsible for formulating response strategies to a human-induced climate change. In 1992 the IPCC was reorganised. WGs II and III were merged into a reorganised WGII, assigned the task of assessing response options in addition to its (continued) assessment of impacts, and a reorganised WGIII assigned the task of assessing the socio-economic dimension of climate change and other cross-cutting issues. All IPCC assessments are to be based on available published literature. At its tenth plenary session the panel decided that it would be reasonable to provide updated assessments roughly every five years. Internally, the bodies constituting the scientific core are principally responsible for the provision of the Assessment Reports, while the WG plenary and the full panel plenary are responsible for accepting and approving the outcome of the work taking place in the scientific core. The panel plenary also decides on the rules of procedure guiding the work of the IPCC.

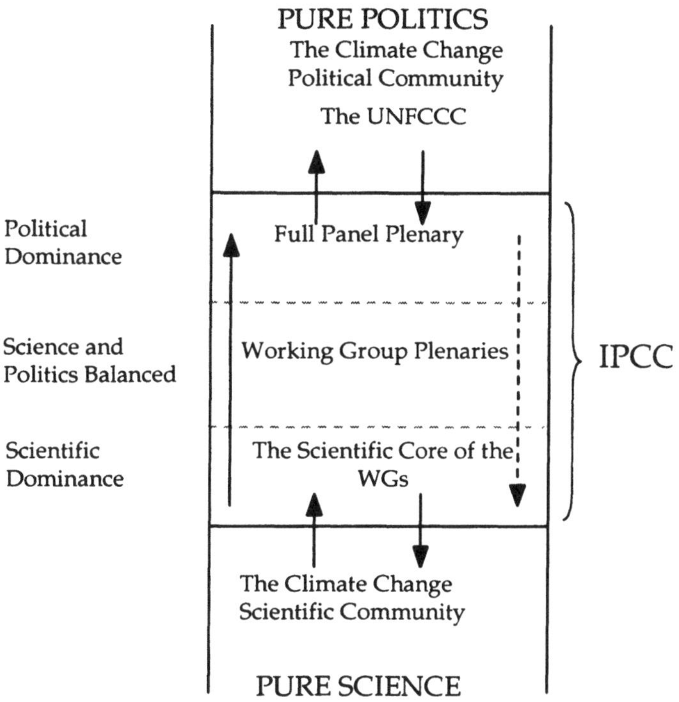

Figure 1. The institutional set-up of the IPCC. The arrows are rough illustrations of formal and informal channels of communication. (See also Lunde, 1991)

The IPCC provides a set of reports differentiated according to how they are approved by the IPCC (see textbox 1). The procedures guiding the panel's adoption of reports were made explicit at the ninth plenary session of the IPCC (June, 1993). Four categories of IPCC material and reports were identified: "Supplementary material" (contributions), "Reports accepted by WGs" (Assessment Reports and Methodologies), "Reports approved by WGs and accepted by the Panel" (Summaries for Policymakers and Executive Summaries) and finally, "Reports approved by the Panel" (Synthesis Reports). An "approved" report has been subjected to detailed, line-by-line discussions in a plenary session of the relevant WG. Larger documents, such as the Assessment Report itself (which usually amounts to several hundred pages), are "accepted" *en bloc* by the plenary, "signifying its view that a report presents a comprehensive, objective and balanced view of the subject matter" (IPCC, 1994; see also report of the ninth plenary session).

The IPCC has published two Assessment Reports, one in 1990 and one in 1995. The panel has also provided several "Special Reports", assessments of a limited range on special issues (1992, 1994). In the provision of Special Reports the rules of procedure guiding the assessment process in general have been followed.

At its eleventh plenary session in 1995, the panel decided that, in addition to the Assessment Reports, the panel should also provide more targeted technical papers on specific issues upon request from the COP (via SBSTA – the Subsidiary Body for Scientific and Technical Advice) or at the panel plenary's own discretion. These reports are to be based on existing assessments, and need not, therefore, be submitted to the extensive acceptance and approval procedures of the s and summaries. This enables the panel to respond more quickly to inquiries from the COP. Technical Papers are to be submitted to expert and government review. The bureau of the panel is intended to function in the role of editorial board in the incorporation of comments received in the review process of Technical Papers (report of the eleventh plenary session, item 5.9).

IPCC reports:

Supplementary material: Scientific contributions to assessments. Subjected to discussions among lead authors and contributors in the scientific core.

Assessment Reports and Methodologies: The full scientific assessment with status as "Reports accepted by WGs". Accepted by WG plenary, but not subjected to detailed discussion.

Summaries for Policymakers and Executive Summaries: Summaries of the full scientific assessment with status as "Reports approved by WGs and accepted by the Panel". Subjected to line-by-line approval by WG plenary. Accepted by full panel plenary, and not subjected to discussion at this decision-making level.

Synthesis Reports: Synthesis of the reports of all WGs, developed by the WG leadership in co-operation with lead authors and specially invited experts with status as "Reports approved by the Panel". Subjected to line-by-line approval by full panel plenary.

Special Reports: Assessments on special issues. Subjected to the review, acceptance and approval procedures of the Assessment Reports in general.

Technical Papers (since 1995): Reports on specific issues, based on existing Assessment Reports, not submitted to the acceptance and approval procedures of the Assessment Reports.

6.3 The Assessment Process of the IPCC

6.3.1 The Assessment Process in Principle

At the start of an assessment process, the leadership of each WG develops a work-plan for the assessment that is to be approved by the plenary of the WG and subsequently accepted by the panel plenary. Teams of lead and contributing authors are nominated by governments. The bureau (chair and vice-chairs) of each WG select lead authors from the lists provided by governments. Contributing authors may also be specially invited, although their participation is considered with due regard to the geographic balance of the groups, particularly with regard to participants from developing countries. Lead authors participate in their personal capacities.

The Assessment Reports are, in principle, to be developed in a series of meetings, in task forces and sub-groups established for particular issues, workshops and conferences, and most importantly, in regular lead- and contributing-author meetings. The summaries of the assessments are also to be developed at this level. The Assessment Reports and summaries are submitted to an extended review process, including both expert and government review. While the "Terms of Reference for the Working Groups", agreed upon in the first plenary session, state that a peer review process should be incorporated in the preparation of the reports of WGs I and II (report of the first plenary session), the extended review procedures were not formally established until the 1993 revision of rules of procedure. The subject of review procedures was discussed at the first session of the IPCC bureau in 1989, when WGI signalled that a review by both contributing and external experts was included in its work plans for the First Assessment (Agrawala, 1998). In 1990, the IPCC bureau confirmed that the draft reports of WGs I and II were to undergo a peer review, although no formal rules regarding who should participate in the peer review or formal mechanisms for a follow-up of the review process were established (Agrawala, 1998). Thus, in the preparation of the First Assessment, the peer review procedure was largely based on a tradition of scientific conduct and trust (Agrawala, 1998). The two-tiered review system was formally adopted at the ninth plenary session of the panel in June 1993.

The 1993 procedures state that lead authors, WG and sub-group chairs and vice-chairs are responsible for incorporating comments from the review "as appropriate". In this regard, lead authors, chairs and vice-chairs are encouraged to arrange wider meetings with principal contributors and reviewers to discuss particular aspects or areas of major differences, as deemed necessary and if time and funding permits (item 2). It is also emphasised in the 1993 rules of procedure that the Assessment Reports "describe different (possibly controversial) scientific or technical views on a subject, particularly if they are relevant to the political debate" (item 4).

The scientific bodies of the three WGs may arrange joint workshops and meetings as deemed necessary for the provision of assessments. Joint workshops and conferences between WGI of the IPCC and the International Ozone Assessment Panel have also been arranged on overlapping topics.

In principle, participants in the scientific core of the IPCC are mainly scientists operating as contributing and lead authors. In practice, there have been significant variations between the WGs in participation patterns of the scientific core (see section 6.3.2). The leadership of the WG (chair and vice-chairs) and (usually) the chairman of the panel also participate at this level.

After the Assessment Report has been revised in accordance with comments brought forward in the review process, final drafts of the assessments and their summaries are to be submitted to the respective WG plenaries for acceptance and approval. While the Assessment Report is accepted *en bloc*, the summaries must go through detailed, line-by-line, discussions.

At WG plenaries, participants function as delegates. Members of the delegations are nominated by national governments. IGOs and NGOs participate as observers but may speak if permitted by the meeting. Finally, lead authors also participate in the WG plenary sessions.

All UN members and members of the IPCC's two sponsoring organisations, the WMO and UNEP, can participate in WG and panel plenaries. Participation at these levels is, therefore, open in principle. It is interesting to note, however, that the IPCC was initially designed on the basis of "core membership" in the WGs. In the report of the first session it is noted that "while participation by all countries interested in the activity of a Working Group would be essential to achieve a comprehensive and balanced outcome of its work, there is a need to provide for a manageable and effective group composition. Therefore, it was agreed that a core membership would be established for each Working Group" (agenda item 3.6). The WGs initially consisted, therefore, of thirteen to seventeen "core members". The inconsistency between this participation procedure and the intergovernmental status of the body was recognised, however, even in the first session in noting that

> this would not imply that other interested countries would not be welcome to join in a Group's activity … on the contrary, ... the Panel … suggested that any country interested in the work of a particular Working Group should be invited to designate a focal point for purposes of communication and *to attend sessions of the Working Group, if desired.* (Report of the first session, agenda item 3.7, emphasis added)

At the second plenary session in 1989, the principle of core membership was abandoned altogether (Report of the second session of the IPCC, agenda item 6.1).

WG plenaries are thus dominated by government officials. Many delegates to WG plenary sessions are, however, scientists working in governmental agencies. Moreover, with the potentially significant role of lead authors at this level, there seems to be a balance between scientific and political dominance (this point is further discussed below).

The Summaries for Policymakers (SPMs) are then to be submitted to the full panel plenary for acceptance. According to the 1993 rules of procedure, the full panel plenary cannot make amendments to a document that has been accepted and approved by a WG plenary. A synthesis of the reports of all the three WGs, prepared by the chairman of the IPCC and the chairs and vice-chairs of the WGs, is also submitted to the panel plenary for approval. Before the draft synthesis report is presented to the panel plenary it has been submitted to a government review process. The synthesis report is subjected to line-by-line discussions before it is approved by the panel.

The panel plenary is also the main decision-making body for the rules of procedure guiding all IPCC activities. Rules of procedure have been developed incrementally, from a very sketchy "terms of reference" guiding IPCC activities in the initial phase (1988–91), to the also rather sketchy rules of procedure adopted at the fifth plenary session (1991), to the more detailed rules of procedure that have guided the preparation, review, acceptance and approval of IPCC reports, the workshop policy of the panel and the tasks and responsibilities of lead authors, contributors and reviewers since the ninth plenary session in 1993.[1]

Inquiries put forth by the COP for scientific information and assessments on particular issues are discussed and decided upon by the plenary. Thus, in addition to the bureau of the panel, the panel plenary is also one main decision-making body for the initiation of special reports and technical papers on specific issues.

Participation in the full panel plenary is dominated by government officials and low-level policymakers. The scientific community is represented at this decision-making level mainly by the leadership of the WGs and the leadership of the panel.

Throughout the IPCC's history, the functioning of the assessment process in practice has varied somewhat in time and especially between WGs. In the next section, we will discuss in more detail the assessment

[1] The IPCC rules of procedure were further revised and formalised in April, 1999. This revision does not fall within the time span of this analysis.

process in practice with a main focus on the proceedings of WGI.[2] To illustrate the significant differences between WGs that have manifested themselves in the assessment process, however, this discussion is contrasted with a brief discussion of the assessment process as it functioned in the old WGIII.

6.3.2 The Assessment Process in Practice: The Proceedings of WGI

The Scientific Core

WGI is perhaps the WG in which the assessment process has best functioned as intended. In many respects, the procedures gradually and incrementally developed throughout the IPCC's history may be regarded as a formalisation of the assessment process as it has functioned in WGI.

In WGI the assessments and their (draft) summaries are developed at workshops, conferences and, most importantly, in the contributor and lead author meetings held at regular intervals during the assessment process. Each chapter of the Assessment Report is penned by teams of lead authors appointed by the bureau of the WG on the basis of government nominations, and is intended to reflect the discussions and scientifically based viewpoints of the contributor and lead author meetings. This work seems generally to reflect the proceedings of a normal scientific process and adheres to the (informal) norms, rules and procedures that govern all scientific endeavour.

Participants thus seem to view these meetings as forums for scientific debate, where participation is regarded as scientifically useful and beneficial in itself. It is, for instance, emphasised by many lead authors that these meetings represent an opportunity to get an update on on-going research and new results within the field at large, and that this opportunity to be at the research frontier constitutes one major motivation for participating (personal communication). The composition of scientists in WGI is cross-disciplinary within the natural science community. In this regard, these meetings also represent a valuable opportunity for participants to develop new contacts and new networks with scientists from other related fields and disciplines (personal communication). Moreover, as pointed out by many lead authors (in personal communication), one main function of the contributor- and lead-author meetings is the opportunity to comment and criticise each other's contributions to the Assessment Report. As put by one lead author of WGI, "we're probably [each other's] most harsh reviewers" (personal communication).

[2] This analysis focuses mainly on the proceedings in WGI because of time constraints and, most importantly, the accessibility of data.

When a draft report has been developed, it is submitted to an extensive, two-tier review procedure that includes both "expert" and "government" reviews. This procedure was, in a sense, introduced by WGI in their work on the First Assessment, when formal procedures for review had not yet been established. The two-tier review procedure was formally adopted in 1993. According to the 1993 procedures, expert reviewers are chosen on the basis of existing scientific networks, proposals from lead authors and the leadership of the WG as well as proposals and nominations from international scientific institutions and IPCC member countries and organisations. The selection of expert reviewers is mainly based on scientific merit (significant publication is one main criterion for the selection of reviewers). Lead authors are mainly responsible for incorporating comments brought forward in the review process. The 1993 rules of procedure specify, moreover, that if time allows, a revised draft should be circulated again for review by experts. In WGI a second review by the experts from the first round is carried out together with the government review.

Governments are encouraged to organise their internal review process so that an integrated set of comments is prepared and made available to lead authors and the leadership of the WG. A final draft, incorporating comments brought forward in the review process "as appropriate" is prepared by lead authors and the leadership of the WG and presented to the WG plenary for acceptance.

Most probably, the leadership of WGI were unaware of the enormous number of comments to be received when this extensive review procedure was first initiated. In an internal letter (dated April, 1990), the co-ordinator of WGI expresses his amazement over 150 *pages* of comments received on the 27-page Policymakers' Summary from about 100 reviewers. Not until the current revision of rules of procedure have formal mechanisms for the handling of these comments been developed.[3]

These two stages of the process – the provision of the assessment and the review process – seem to be well shielded against purely politically motivated attempts to influence the substantive conclusions of the report. Overall, participation by non-experts is rare within the scientific core of WGI. To the extent that non-experts do participate at this level, the sheer scientific complexity of the issue effectively prevents undue political influence. Moreover, none of the interviewed lead authors have experienced political pressures at this stage of the WGI assessment process. Rather, the most important pressure and constraint as experienced by lead authors is the

[3] The current revision establishes an editorial review process where one or two review editors, who have not been involved in the development of the assessment and its summaries, are selected for each of the chapters of the assessment and each of the summaries of each WG report (Decision Paper from the thirteenth panel plenary session in 1997).

time constraint which reportedly often has implied an inadequate handling of comments brought forward in the review process (personal communication; see also report of the eighth panel plenary session).

This does not imply, however, that the proceedings in the scientific core of WGI have been entirely non-controversial and harmonious. Internal letters and communications within WGI (mostly covering the period until the 1992 report) witness a sharp scientific debate in the assessment process. These controversies seem, however, mainly to have been resolved through debates and discussions whereby compromises reflecting major viewpoints are developed[4] or by developing generous intervals and ranges of uncertainty (personal communication). The prime example of this is the suggested interval of a 1.5–4.5°C increase in the global mean temperature at equilibrium resulting from a doubling of CO_2 concentrations in the atmosphere (the climate sensitivity). While some modellers would maintain that the temperature increase is more likely to be 4.5°C than 1.5°C and others argue vice versa, they all agree that the temperature increase lies within this range.

Also, in chapter 5 of the first Assessment Report, a system to indicate level of confidence was developed where "the number of *'s indicates the degree of confidence determined subjectively from the amount of agreement between models, our understanding of the model results and our confidence in the representation of the relevant process in the model. Five *'s indicate virtual certainties, one * indicates low confidence" (Mitchell et al., 1990: 135).

The scientific core does not operate under a consensus rule. As pointed out above, the 1993 rules of procedure specifically emphasise that different and possibly controversial scientific views on a subject are addressed and described (item 4). The question of scientific consensus itself is, however, ambiguous among scientists operating within the scientific core of WGI. While some lead authors maintain that consensus is not, and should not be, a requirement in the IPCC process, others see the development of consensus as one objective of the process (personal communication). Many of the lead authors interviewed seem to be uncomfortable with the notion of consensus and unclear about exactly what it entails. However, whether or not this is considered a requirement of the process, they all seem to agree that the desire to achieve some sort of consensus has not seriously restricted the scientific debates and discussions that have taken place within the scientific core, although some do maintain that, at least in its initial phase (the first assessment), the IPCC process has not been sufficiently inclusive.

[4] See also the discussion in section 7.4.4 on the specific case of the conflict between WGs I and II on the use of the paleo-analogue method as a forecasting technique.

Representatives of the leadership of the WG and the panel, on the other hand, do not consider the development of consensus, understood as the formulation of conclusions and propositions supported by the entire scientific community, as a task for the IPCC. Consensus understood as the development of an assessment which reflects the scientific debate *with* its inherent uncertainties and controversies and which, hence, is a consensual *representation* of state-of-the-art knowledge, on the other hand, is considered an important objective of the IPCC process by the IPCC leadership (personal communication).

The WGI Plenary

The revised draft of the assessment and its summaries are then submitted to the WGI plenary for acceptance and approval. At this level, the discussion takes on quite a different character. While the full scientific Assessment Report is accepted by the plenary *en bloc* and usually without further ado, the summaries – the Executive Summary (ES) and the Summary for Policymakers (SPM) – undergo a detailed and time consuming revision where the formulations of the documents are discussed and negotiated, line-by-line, during the course of the meeting.

WGI plenaries are largely made up of national delegations, comprising government officials, low-level policymakers and/or scientists with governmental affiliations. National governments may also to a varying extent send independent scientists as members of national delegations to WGI plenary meetings.

Scientists are represented at this decision-making level mainly by representatives of the teams of lead authors of each chapter of the assessment. Lead authors have acquired a special status as authorities in the debate, and substantive changes to the text of the summaries are not made without consent from the lead authors of the chapter in question. Thus, while scientists may be outnumbered by government officials at this level, they still have a significant amount of control over the documents.

Scientific, environmental and industrial IGOs and NGOs participate as observers to WGI plenary meetings and may speak as permitted by the meeting.

IPCC WG plenaries operate under a decision rule of consensus. The 1991 rules of procedure states, "[i]n taking decisions, drawing conclusions, and adopting reports, the IPCC Plenary and Working Groups shall use all best endeavours to reach consensus." Furthermore, in the 1991 rules of procedure it is stated that "[i]f consensus is judged by the relevant body not possible...for conclusions and adoption of reports, differing views shall be explained and, upon request, recorded." ("Principles governing IPCC work" from 1991, item 6).

The consensus approval of WGI summaries takes place in a highly politicised and polarised environment. The documents are exposed to efforts of both strengthening and watering down their substantive conclusions. The political motivations behind many of the suggested text changes are usually quite obvious. For instance, while the small island states (who might suffer substantially by sea level rise), supported by environmental NGOs, often push for stronger language on climate change risks and the need for immediate action, the major oil-producing countries (who would suffer a substantial economic loss from a shift away from fossil fuels), supported by the industrial lobby, often advocate formulations emphasising scientific uncertainty and greenhouse gases other than CO_2 (see also Agrawala, 1998). In between are the lead authors who try to draft and redraft the summaries in a balanced and scientifically defensible language.

The WG plenaries, therefore, are in many respects a peculiar performance, illustratively described by Agrawala as "a fox-trot performed by a drunken couple: one lurch forward, followed by a sideways stagger, then a stumble backwards" (Agrawala, 1998). It seems, however, that the final product is usually not very different in substance from the drafts. First, the polarisation in the attacks themselves contributes to a balance, since they usually somehow counteract each other. If efforts by the oil-producing countries (Kuwait and Saudi Arabia in particular) to water down or shift emphasis in the text are not met by the group of small island states (AOSIS), the USA very often adopts a role as counterweight to these proposals.[5] The balance is also, however, the achievement of the lead authors and their persistent insistence that the formulations must be scientifically substantiated and correspond to the conclusions of the bulk report. The scientific authority of lead authors in this debate also seems to be respected by most delegations, and lead authors do get the final say in these discussions.[6] This is, moreover, critical for the scientific credibility and authority of the document.

[5] This was for instance clearly illustrated at the fifth plenary session of WGI in Madrid when a Saudi Arabian proposal to restrict the climate sensitivity to the lower end of the range was met by the USA with a counterproposal to restrict the sensitivity range to the upper end of the range, i.e., 3.5–4.5°C (personal observation).

[6] For example, at the Madrid meeting of WGI there was for instance a lengthy discussion on the formulation, "More convincing recent evidence for the attribution of a human effect on climate is emerging from pattern-based studies,...". Kuwait and Saudi Arabia wanted the formulation to be either "Some preliminary evidence..." or "More convincing, but preliminary, evidence...". On this occasion the Chairman of the panel, Bert Bolin, ruled that given the unanimous support by all lead authors present for the original formulation, the conflict should be resolved by recording the dissenting views of these two delegations in a footnote.

The end result then, is a report with carefully hedged language usually approved by consensus by the WG plenary.[7] The development of consensus in the WG plenary may take at least two routes. First, the procedure encouraged by the leadership of the WG is to move discussions on particularly controversial issues out of the plenary to side meetings with fewer participants. When this procedure works, the side meeting develops a revised text on the sections in question, which is then merely re-inserted in the document by the plenary. In the development of the 1994 WGI report, this procedure worked very well at the WGI plenary session in Maastricht.

In some cases, however, delegations may choose not to acknowledge the text developed in side meetings and reopen disputes, previously settled in side meetings, in the plenary. In such cases, the WG plenary revision of the summaries becomes an extremely time consuming, exhausting and intensely political process. This was the case in WGI's plenary approval of the SPM to the Second Assessment Report in Madrid 1995, where most efforts to channel disputes and controversies to smaller side meetings were effectively stopped by the major oil-producing countries. They simply did not accept a procedure whereby side meetings operated in parallel with plenary discussions.[8] Even in this politically contentious meeting, however, the relatively strong statement that "the balance of evidence suggests that there is a discernible human influence on global climate" was agreed upon, although it took three days with intense negotiations. Agreement on this sentence was by far the most controversial and difficult issue on the agenda of this meeting. Therefore, it is noteworthy that the corresponding sentence of the draft report, as it was formulated *before* the negotiations of the 1995 WGI plenary started, reads: "Taken together, these results point towards a detectable human influence on global climate." Even the most positively minded may ask, What is the point of this exhaustive and time-consuming exercise?

[7] On one occasion, the approval of the SPM to the Second Assessment Report, Kuwait and Saudi Arabia requested that their dissenting view regarding a formulation in the report be recorded in a footnote. Due to some confusion over whether or not dissenting countries should be named or not, this question was left open for decision at the following panel plenary, at which the request was withdrawn (see also footnote 6).

[8] One may suspect that this was a conscious delaying strategy of these delegations since they usually voiced their objections to the procedure of side meetings after, rather than before the event. Hence, side meetings had already worked out draft proposals to be inserted in the text, when these delegations voiced their objections and reopened the issue. The result was that the meeting did not have the time to go through the whole document intended to be the SPM. This situation was resolved by converting the document intended to be the Executive Summary, which had been approved by the meeting, to an SPM. The document originally intended to be the SPM was subsequently accepted by the *panel* plenary as a "Technical Summary" to the Assessment Report of WGI.

Essentially, the point seems to be that the WG plenary discussions represent the first step towards acquiring a political acceptance of the knowledge base and its substantive conclusions. Having undergone this thorough and detailed treatment, where alternative formulations and interpretations of the corresponding formulations in the bulk report have been discussed and negotiated, the substantive conclusions of the knowledge base are in a sense "tried out" and "digested" by policymakers. Having survived this intense scientific and political scrutiny with their scientific credibility and authority intact, the substantive conclusions come out as more robust and are not easily deconstructed.

The Panel Plenary

The accepted and approved Assessment Report and summaries are then submitted to the full panel plenary for acceptance. A report that has been accepted or approved by the WG plenary cannot, however, be amended by the full panel plenary. This institutional device, formally established in the 1993 revision of the IPCC rules of procedure, is important for ensuring consistency between the summaries and the Assessment Report upon which the summaries are based. As discussed above, the lead authors' scientific authority is used at WGI plenary meetings to ensure this consistency and also to prevent scientifically unsubstantiated formulations from entering the summaries. While lead authors usually participate at the WG plenary level, they usually do not participate in the full panel plenary meetings. This institutional device also prevents the reopening in the full panel plenary of controversial issues already settled in the WG plenaries. A situation as the one that arose in the 1990 plenary session with the adoption of the summary of the First Assessment Report[9] may thus be avoided. In this case, negotiations in the panel plenary almost went into deadlock and threatened to overthrow the whole process (see for instance, *New Scientist*, 8 September 1990). At the time, explicit rules of procedure for the panel's adoption of summaries were not developed. A summary based on the three reports, developed by Bolin himself, was met by massive criticism and brought latent conflicts into the open.

Agreement was achieved in this situation by using text already agreed upon in the WGs. While Bolin ascribes this situation to his own inexperience and naïvety (personal communication), new, and most important, explicit, rules of procedure have contributed to a smoother decision-making process.

[9] The provision of a "Synthesis Report" based on the SPMs of all the three WGs was only decided after the provision of the First Assessment Report, in preparation of the Second Assessment. At the time, therefore, there was no established procedure whereby to develop a "synthesised" summary of the three SPMs.

Thus, while the assessment process is formally finalised with the acceptance of the assessments and summaries by the panel plenary, it is in practice finalised with the acceptance and approval of the reports by the WG plenary (according to the 1993 rules of procedure).

The panel plenary approves the Synthesis Report drafted by the leadership of each of the three WGs in co-operation with a specially invited group of scientists, lead authors and experts. The 1995 Synthesis Report was developed and discussed at several conferences with broad participation. The procedure by which consensus on the Synthesis Report is developed in the panel plenary is, in form, similar to the negotiations taking place in WGI plenary meetings. A notable exception is the near absence of scientists at this decision-making level. This places a special burden and challenge on the members of the drafting team that are present and the scientific leadership of the WGs and the panel. In the development of the Synthesis Report to the Second Assessment, conflicts over formulations were resolved by resorting to agreed-upon language drawn from the SPMs. A more frequent recording of dissenting views as well as defining and qualifying statements in footnotes were also employed as tools for the development of consensus.

6.3.3 The Assessment Process in Practice – In Contrast to the Proceedings of the Old WGIII

The differences between the proceedings of WGI and the old WGIII are significant. Before 1992, WGIII of the IPCC was assigned the task of "Formulating response strategies" to a human-induced climate change. The old WGIII did not, therefore, conduct an assessment of existing and published literature within a specific field of expertise, as did WGs I and II. Rather, their task was more one of formulating more or less explicit advice, a task which was very closely associated with the politics of a human-induced climate change. Even more so, since the submission of their report took place simultaneously with the development of the knowledge base from which this advice was to be drawn, coupled with the fact that a formally established political forum for climate change negotiations did not exist until the establishment of the Intergovernmental Negotiating Committee for a Framework Convention on Climate Change (INC) in 1991.

Given the nature of their task, the old WGIII could not draw upon a clearly defined scientific network or field of expertise in the provision of their report. The WG became dominated by (low-level) policymakers and negotiators at all decision-making levels, with only a few independent legal and environmental experts, and the standard procedures for scientific peer review did not apply. Thus, the work of the old WGIII was not peer reviewed, although the Summary for Policymakers was submitted to a government review. The result was that the old WGIII experienced a significant politicisation of their work.

As emphasised above, there were practically no formally established rules of procedure guiding the work of the IPCC bodies during this period. When the task of the WG was firmly associated with a clearly defined scientific community, as was the case in WGI, this implied that the work could proceed in accordance with the rules and norms guiding scientific endeavour in general. When the task could *not* be clearly identified in terms of a coherent and well-organised scientific community, however, as was the case in the old WGIII, this lack of formal rules implied that there were *no* guiding principles for how the work was to be carried out. Thus, the lack of formal rules implied, in the case of the old WGIII, that there were no instruments by which to handle the very strong political pressures directed at the WG during this phase. Rather, with informal rules as the only guide, the informal rules of politics were more natural guides to the work of the old WGIII than the informal rules of science. Also, the lack of formally established political bodies, external to the IPCC process, to which strictly political issues could be redirected, seems to have reinforced the vulnerability of this WG to a substantial politicisation. In this phase, policymakers and governments were very uncertain about what the IPCC process might lead to and which (political) implications the report of WGIII might have. Without a proper (political) forum where these political aspects could be discussed, they were unwilling to lose control over the work and proceedings of this WG.

The old WGIII, therefore, became the "cockpit" for climate change politics in the pre-1991 period (Brenton, 1994:179), and its most important function was to serve as a forum for pre-negotiations (see also Boehmer-Christiansen, 1993; Agrawala, 1998). This was, by all means, a valuable and necessary contribution at this stage of the process, but it was *not* a scientific task. The political "infiltration" of the old WGIII seems to have generated some uncertainty regarding how their work was to be evaluated: as a political or as a scientific document. Within WGI and the leadership of the panel in general, moreover, there seems to have been some concern that this ambiguity could have significant credibility implications for the whole process (personal communication). The IPCC Chairman stated, for instance, that with the establishment of the INC (1991), the IPCC would be better able to draw a distinction between the scientific/technical and the political parts of the tasks ascribed to the panel, and that "it is of great importance, ... , that the IPCC task be limited to scientific/technical assessments in order to allow the compilation of as objective information as possible to serve the political negotiation process".[10] In this regard, the establishment of a negotiating committee enabled the IPCC to reorganise itself, withdraw from the (explicit) advisory function and reformulate its task to a provision of *assessments* for all WGs.

One of the areas in which the politicisation of the old WGIII represented a major problem was in the development of the emission scenarios. At its first session WGIII decided to develop global emission scenarios that would serve as "(1) an initial reference and guidance for the work of the subgroups of [WGIII] (...), and (2) a first basis for the work of Working Group I (...) and Working Group II (...)" (IPCC, WGIII, 1990 report: 17). Scenarios of future emissions were developed corresponding to (1) the equivalent of a CO_2 doubling from pre-industrial levels by about the year 2030; (2) a doubling by approximately 2060; and (3) a doubling by about 2090 with stabilisation thereafter. Thus, the scenarios were developed in a top-down approach, by assuming the equivalent of a CO_2 doubling of atmospheric concentrations at a certain time instead of a point of departure in current emission structures and economic assumptions about likely developments in future emissions. An internal letter indicates that this approach generated some concern in WGI:

> ...the scenarios were generated in a very unorthodox way and, the more I think about it, the more I find it hard to take them seriously. The basic problem stems from the dictat [sic] handed down from the Washington WG3 meeting, that, instead of generating scenarios for each gas appropriate to three economics-based estimates of emissions from the whole range of sources (...), the starting points were to be an effective doubling of CO_2 concentrations by 2030, 2060 and 2090.
>
> Using these starting points, EPA have projected CO_2 emissions economically using a high and low GNP growth assumption, and then adjusted emissions of the other greenhouse gases so that the effective doubling of CO_2 occurs in these years for both growth assumptions. The high and low growth scenarios for all greenhouse gases are then averaged to give the single scenario for each of the three years to effective $2xCO_2$. The main weakness in this method lies in the adjustment of the other greenhouse gases; whilst EPA claim that the emissions arising from this adjustment does not go outside the bounds of impossibility, neither do they follow any sort of logic. Certainly they cannot be traced to any economic assumptions. ...

[10] IPCC/TF/1st/Doc.2: "An analysis of principles and a proposal for structural changes of the IPCC", submitted by the IPCC Chairman for discussions at the first session of the IPCC Task Force on IPCC Structure, February 1992, emphasis deleted.

> [These scenarios] will be of no use whatsoever to policy deliberations. What can be done with a climate scenario based on an effective 2xCO_2 date of 2030, when there is no direct link to any emission assumptions? ... I do wonder whether by ducking the issue of real emissions scenarios now we may be storing up trouble for later. (Internal letter of 2 May 1989, from WGI co-ordinator Geoff Jenkins to Alan Apling, Head of Science Unit at the UK Department of the Environment)

Given the level of politicisation of the old WGIII, the point of these emissions scenarios, for many delegations, was probably precisely their lack of policy relevance. This is perhaps one example of how the politicisation of the old WGIII restricted the options of the WG to conduct a scientifically "sound" estimation.

These politicisation problems in the development of the 1990-scenarios were recognised by the IPCC Task Force on IPCC Structure, and the task of developing emission scenarios was transferred to WGI. The 1992 scenarios developed by WGI were then evaluated by (the new) WGIII in the 1994 Special Report.

The 1992 restructuring meant that the task of the old WGIII was assigned to the restructured WGII and reformulated to become an assessment of response *options*. The new WGIII was assigned the task of providing "technical assessments of the socio-economics of impacts, adaptation and mitigation of climate change over both the short and the long term and the regional and global levels" (report of the eighth plenary session, item 6.10). In contrast to the situation before 1992, all IPCC WGs are now providing assessments on the basis of published literature with scientific quality control (peer reviewed literature). Thus, as of 1992, the IPCC does not formulate policy recommendations. A peer review process is, furthermore, formally established for all the WGs (report of the eighth plenary session, item 6.10.2).

The mandate of the new WGIII lies within the sphere of socio-economics. Hence, as opposed to the situation before 1992, the task of the WG is now firmly associated with a clearly defined scientific community or research network, and the WG now functions as the other WGs of the IPCC. Economics is, however, more closely linked to value judgements than natural science, and this may, in itself, lead to confrontations between scientists and policymakers. This is for instance illustrated in the controversy regarding the value of a "statistical life" which took place in relation to the development of the Second Assessment Report in 1995, and which developed into a heated debate between (some) lead authors and developing-country delegates (see, e.g., *Nature*, vol.378, 1996; vol. 379, 1996; *GECR*, 11 August and 8 September 1995). Although it is emphasised in the report of WGIII that what is valued is a *statistical* life – "a change in the risk of death" – not human life itself (Pierce et. al., 1995: 196), the significant difference in the value assigned a statistical life in OECD as compared to developing countries is controversial.[11] Attributing a monetary value to statistical lives is controversial even among economists and is obviously problematic within the framework of a process so heavily marked by a potentially paralysing North–South controversy. The strength of this controversy may, however, also have been a result of an inadequate incorporation of review comments. It is possible, therefore, that this controversy could have been avoided, or at least minimised, had the issue been dealt with a bit more cautiously in the first place. Moreover, an external agent supervising the incorporation of review comments might have been instrumental for avoiding the situation that arose in WGIII (see also Agrawala, 1998). The controversy was resolved at a resumed WGIII plenary meeting in broad consultations between delegations and the lead authors of the chapter in question.

6.4 The Capacity of the Institutional Arrangements of the IPCC to Serve the Four Main Functions

Chapter 4 suggested that institutional design may enhance the effectiveness of the science–policy dialogue to the extent that the institutional apparatus serves four main functions: maintaining the scientific autonomy of the scientific bodies; ensuring some level of science–policy involvement; ensuring a geographic balance in participation in the scientific bodies; and providing mechanisms for conflict resolution. In this section, the capacity of the institutional apparatus of the IPCC to serve these functions is investigated.

6.4.1 Scientific Autonomy

Maintenance of scientific autonomy is understood and operationalised as a function of the following set of variables: the main function of the body; who appoints and funds participants; the criteria employed in the recruitment of scientists; and the operational autonomy and unity of the scientific body. As discussed above, there are significant differences in the role and functioning of the IPCC's three decision-making levels. Thus, these variables will be discussed in terms of each decision-making level.[12]

In general, the scientific core of WGI has a relatively high score on the autonomy dimension: The main function of this body is to provide scientific

[11] While a cash value of $1.5 million was assigned to a statistical life in the OECD, the similar measure for developing countries was a mere $150,000 (Agrawala, 1998).

[12] Note that the analysis is mainly focused on the proceedings of WGI.

assessments on the basis of published scientific literature. Participants are largely scientists active in research, recruited on the basis of scholarly merit, appointed by the scientific leadership of the WG and the panel, and funded through the regular channels for scientific funding at the national level. Due to the lack of formal procedures during the initial phase of the IPCC process, the operational autonomy of the scientific core of WGI has been high. This operational autonomy has been maintained also after the IPCC formalised their rules of procedure. The scientific core is free to organise its own work and allocate specific roles. The selection of lead authors is, for instance, done at this level. Governments do nominate candidates for this role and the WGs are required to take the geographic balance of the teams of lead authors into account in their selection of lead authors. Governments do not, however, have a direct influence on the actual choice of lead authors. The lack of formal rules during the initial phase of the process implied that WGI could carry out their work in accordance with the informal rules and norms governing all scientific endeavour. Thus, the formalisation of the procedures employed by WGI during the initial phase of the process as guides to all IPCC activities implies a large degree of correspondence between the formalised rules and procedures of the IPCC and the informal norms and principles of scientific research in general.

With regard to the variable we have referred to as the "unity" of the scientific body (see chapter 4), the picture is mixed: The IPCC as a whole has an independent institutional basis, in the sense of it being a separate body from the negotiating/regulatory body. Both the formal and the operational autonomy of the IPCC in relation to the COP are high. The two bodies are organised under different UN bodies, and the COP is in no position to place either restrictions or prescriptions on the manner in which the IPCC chooses to organise its work. The COP may specify which questions they wish the IPCC to address, but the IPCC is free to choose whether and when to respond to the inquiry. On the other hand, not all members of the IPCC (at all decision-making levels) are scientists. Within the scientific core of WGI, however, (practically) all members are scientists.

Moving to the plenary level of WGI, the picture changes significantly. In contrast to the situation within the scientific core, the WG plenary is characterised by a relatively low scientific autonomy. Participants are appointed and funded by national governments and recruited largely on the basis of their administrative position in the national bureaucracy, although some governments appoint independent scientists as members of their national delegations. The meeting and administrative expenditures of the IPCC are financed by national governments on the basis of voluntary donations. The IPCC also receives some financial support from UNEP and WMO.

The operational autonomy of the WG plenary is high in relation to the negotiating/regulatory body, but delegates to the WG plenary do operate under the instruction and control of national governments. The main function of the WG plenary is to accept and approve the assessments presented by scientists. While this function clearly cannot be defined as doing research, they are not exactly primarily concerned with policy implications either, although many *delegations* may have the policy implications of the knowledge base as their prime concern. The inference of policy implications is not, however, a direct outcome of the process at this decision-making level. In this regard we may categorise the main function of the WGI plenary as moderately political. Moreover, the function of the WGI plenary is more closely related to the transformation of scientific knowledge into decision premises than to the provision of the knowledge base itself. As was the case with regard to the scientific core of WGI, the "unity" of the WGI plenary is mixed: As part of the IPCC it has an independent institutional basis, but participation includes professions other than scientists. The score of the WGI plenary along these variables, therefore, indicates a low level of scientific autonomy. It should be emphasised, however, that this score is to some extent counterbalanced by the role and participation of lead authors at this decision-making level. Recalling the very central role of lead authors at WGI plenaries in the adoption of reports, this point is important and modifies the overall score towards a mediate level of autonomy.

The score of the full panel plenary follows the score of the WGI plenary when looking at participating delegations. The important difference at this level, however, is the near absence of scientists, which overall implies a low score on scientific autonomy dimension.

Overall, therefore, the autonomy of the various bodies of the IPCC seems to have varied somewhat, both vertically, between decision-making levels (between plenary, WG and scientific core), and horizontally, between the scientific cores of the three WGs. As the discussion shows, the main distinction between the bodies runs between the more politicised WG and panel plenaries and the scientific core constituted by task forces, subgroups and the lead- and contributing-author meetings. While the scientific core is characterised by a relatively high level of scientific autonomy, the WG plenary and the full panel plenary are characterised by a mediate and a low level of scientific autonomy (respectively). Although the decision-making authority of the IPCC lies in the panel plenary, where the intergovernmental status of the institution has significant impact on the composition of participants, this does not seem to have severely restricted the autonomy and integrity of the scientific core of WGI.

In contrast to WGI, the old WGIII is characterised by governmental control in all respects and at all decision-making levels, and hence a low degree of scientific autonomy. It is interesting to note the significantly different functioning of two WGs operating under the same (lack of) formal rules and procedures. The key to understanding this difference is the nature of the task and hence the composition of participants to the WGs. The task assigned to WGIII before 1992 lies very close to the political domain and it is not encompassed by any clearly identifiable scientific discipline or group of experts with pre-established scientific networks. Without scientists, an associated scientific community/network and the norms and rules of scientific endeavour as an informal guide, it was impossible to maintain the scientific autonomy and integrity of the old WGIII. Due to these circumstances the provision of the report was moved from the bodies constituting the "scientific core" to the even more politicised decision-making levels of the IPCC (WG and panel plenaries), and the group's autonomy was reduced even further. Even after 1992, the nature of the task assigned to WGIII is such that the distinction between science and politics is blurred. On the other hand, in their balancing between science and politics, the entire IPCC is "walking on eggs" (expressed by Michael Zammit Cutajar in personal communication).

6.4.2 Science–Policy Integration

The capacity of the institutional apparatus to ensure communication and involvement between scientists and policymakers is understood as a function of: the establishment of formal links between scientific and decision-making (regulatory) bodies; the extent to which the tasks of research and advice are functionally differentiated; and the extent to which arenas for an interactive dialogue between scientists and policymakers are provided.

The IPCC reports to UNEP and WMO, while negotiations, by implication of UN Resolution 45/212, are under the control of the UN General Assembly. The resolution acknowledges "the important work accomplished by the [IPCC]" and the ad hoc secretariat of the INC is requested "to co-operate closely with the [IPCC] to ensure that the panel can respond to the needs and requests for objective scientific and technical advice made during the negotiation process". During the initial phase of the IPCC process, a negotiating body was not yet established. During this phase WGIII seems to have served the function of an informal policy arena.

In general, there seems to have been relatively few formal channels for communication between the IPCC and the INC before 1993. During this phase a very important communicative function was served by the leadership of each of the bodies, through representation: The chairman of the IPCC was invited to speak at INC sessions reporting on the progress and findings of the panel, and the chairman of the INC was invited, in an *ex-officio* capacity, to attend the IPCC (plenary), the bureau and the working group sessions. Overlapping participation seems, moreover, to have played an important role in strengthening the relationship between the bodies, especially during this phase, since many of the participants in the IPCC plenary sessions were also delegates at INC. sessions, especially the delegation heads of the IPCC plenaries.[13] Otherwise, the IPCC mainly communicated through their provision of assessments and, especially, the summaries for policymakers.

In March 1993, these rather informal channels for communication and dialogue were supplemented by Bolin's initiative to establish the "IPCC-INC Joint Working Party" to improve the dialogue between the two institutions. This group has met since November 1993, after 1995 under the name of the "Joint Working Group", and consists of the senior officials of the two institutions[14] (Agrawala, 1998; personal communication with Bert Bolin). Furthermore, the decision at the eleventh plenary session of the IPCC to provide more targeted Special Reports and Technical Papers on specific issues as requested by the COP also contributes to improving the dialogue between the two bodies.

Thus, both formal and informal channels for communication and dialogue between the IPCC and the INC/COP are established, although these channels seem to have been incrementally developed during the course of the process. Dissatisfaction with the IPCC's ability to accommodate their provision of reports to the time schedules of the Convention has, however, been expressed (*New Scientist*, June 1994). This complaint has to some extent been met by the decision to provide Technical Papers that address the questions and needs expressed by the COP. While these links between the IPCC and the INC/COP do suggest at least a moderate level of involvement between the two institutions, the relationship between scientists and policymakers within the IPCC itself seems to be more important for the overall score of the IPCC along this dimension.

The task of formulating advice was part of the IPCC's agenda only during a short phase of the IPCC's history (formally until 1992, in practice only until 1990). During this period, the provision of assessments was clearly separated from the formulation of advice. The two tasks were conducted by different bodies within the IPCC (WGs I and II made assessments and the old WGIII was responsible for the formulation of advice), and scientists making the assessments were not involved in the formulation of advice. On the other hand, while being clearly separated

[13] During the course of the process, the IPCC and the COP have become more specialised, and the tendency towards overlapping participation between these two bodies has been reduced.

[14] See also Discussion Paper prepared by Bert Bolin for the tenth plenary session in 1994, "The Future of the IPCC".

within the IPCC, both functions were served by the IPCC during this period and were in that sense closely linked, although the IPCC has been criticised for a lack of co-ordination between these WGs. However, due to the circumstances discussed above, the old WGIII may be regarded as a largely political forum for pre-negotiations, and the texts drafted by this group held general, compromise formulations bearing clear signs of having been negotiated. Thus, the "advice" was based as much on purely political evaluations as scientific. The work of the old WGIII has, however, been argued to have made an important substantial contribution to the early negotiations in the INC, in terms of providing a "preliminary 'roughing-out' of possible options" (Borione and Ripert, 1994; see also Agrawala, 1998).

With the restructuring in 1992, the advisory function of the IPCC has dwindled. Still, the main purpose of the IPCC being to inform policymakers, their work does imply some sort of implicit advice, and over the years they have developed more sophisticated methods for separating science and politics in this regard. This is perhaps best illustrated in the 1995 Synthesis Report developed to aid policymakers in making judgements about what constitutes a "dangerous anthropogenic interference with the climate system" (formulation of Art. 2 in the FCCC). In this report, alternative paths to achieving a stabilisation of CO_2 concentrations in the atmosphere at various levels (450, 550, 750 and 1000 parts per million by volume) are illustrated. In this case the CO_2 emissions reductions needed to stabilise concentrations at these various levels are illustrated without having to rely on arbitrary speculations about future energy use, population growth etc., as was very much the case with regard to the development of scenarios.

Already in 1990, *New Scientist* reported that the scientists of the IPCC had learned "how to talk science to politicians" while at the same time "not to play at being politicians" (Comment, *New Scientist*, 8 September 1990). While this statement may have been somewhat premature in 1990, it certainly seems to have gained validity in the period since. There are examples where the formulations of the WGI reports have been carefully worded with due consideration of the distinction between science and politics. This seems, for instance, to have been the case in the 1992 Supplementary Report, where WGI discussed the question of how their findings on the negative warming potential of SO_2 should be presented because of the possible policy implications (personal communication with participants). A similar discussion on the boundaries between science and politics also took place at the WGI plenary in Maastricht in 1994 in the context of a discussion on the new global warming potential (GWP) for methane. In this case the appropriate boundaries between science and politics were not at all obvious and were subject to discussions between the lead authors of the WG and scientists with close links to policymaking (see

Shackley and Skodvin, 1995). The WGI leadership also seems to view their own work as one of providing some sort of advice to policymakers. This is particularly indicated in the internal correspondence in relation to the conflict between WGs I and II over the paleo-analogue method (see section 6.4.4).

Thus, while the provision of assessments and the formulation of advice have been formally separated, the IPCC is characterised by close links between scientists and policymakers, especially in WG and panel plenaries. This indicates a moderate to high level of involvement at these decision-making levels of the IPCC.

The intergovernmental status of the IPCC implies, as our discussion shows, an increased vulnerability to politically motivated influence on the substantive content of IPCC reports (especially the SPMs and Synthesis Reports). The approval of IPCC reports by the WGI plenary has often become time-consuming negotiations clearly marked by politics and strategic thinking. This mode of developing a knowledge base has also provided IPCC sceptics with sharp ammunition in their criticism of the IPCC. It is, therefore, important to recognise what has been gained by this mode of operation. The WGI plenary serves an important function as an arena for an interactive dialogue between scientists and policymakers where the interpretation of scientific findings and the appropriate limits to what the IPCC can and should engage in can be discussed (including the boundaries between science and politics). This potential of the WGI plenary in this respect is particularly well illustrated at the fourth WGI plenary session in Maastricht, where the contribution by WGI to the 1994 Special Report was discussed, accepted and approved. At this session, smaller side group meetings were established to revise and discuss the various chapters of the executive summary and the SPM. These side group meetings were characterised by scientific discussions between the scientists of WGI and scientists and experts with governmental affiliations as well as low-level policymakers. In these discussions, the element of strategic political thinking, which is very much present in some WGI plenary sessions, was less obvious (personal observation). It is also important to note, however, that the availability of this instrument strongly depends on the level of conflict characterising the discussions in the plenary. As we have seen, the fifth plenary session in Madrid was too conflictual and difficult for the employment of side meetings to function very well. In general, however, given their devotion of three to five whole days (and sometimes nights) to discussing the appropriate formulations by which to communicate the findings of the Assessment Reports taken into account, the WG plenaries can certainly be characterised as arenas for an interactive science–policy dialogue.

Thus, the IPCC has served to provide arenas for an interactive dialogue between scientists and policymakers that are clearly separated from the bodies actually developing the assessments – above all in WG and panel plenaries, and also, more informally, at smaller side meetings to WGI plenary sessions. This indicates a moderate to high level of involvement between scientists and policymakers.

Overall, this discussion suggests that the IPCC is characterised by a relatively high level of involvement between scientists and policymakers, to some extent as a function of the relationship between the IPCC and the INC/COP, but above all as a function of the close links between scientists and policymakers within the IPCC itself. It is worth noting, however, that it is first and foremost the design and functioning of the WG plenaries and the full panel plenaries that contribute to this effect – that is, the same decision-making levels for which we noted a relatively low level of scientific autonomy. Moreover, it is also worth noting that "involvement" in this process mainly refers to the involvement of policymakers in the assessment process. Scientists seem to a very little extent to have been actively involved in the policymaking process (apart from their participation at WGI plenaries to the extent that WG plenaries can be seen as part of the policymaking process).

6.4.3 Geographic Representativeness

The necessity of a geographic representativeness within the IPCC was recognised early. In his announcement of the proposal to establish the IPCC of March 1988, WMO Secretary General G.O.P. Obasi noted that membership of the panel should, among other things, ensure equitable geographic representation (Agrawala, 1998b). In their effort to keep the IPCC a small forum, however, the IPCC was designed on the basis of "core membership" in the WGs. This design was abandoned in 1989 to allow for a more widespread participation, especially by developing countries. In this regard, Bolin commented that "right now, many countries, especially developing countries, simply do not trust assessments in which their scientists and policymakers have not participated" (Schneider, 1991: 25, cited in Agrawala, 1998b). Thus, the dominance by developed, Western, industrialised countries in IPCC bodies, especially during the initial phase of the process, was recognised as a major challenge to the success of the process. In 1989, following the recommendation of an ad hoc group on developing-country participation established by the Bureau early in 1989 and chaired by Saudi Arabia, the Special Committee on the Participation of Developing Countries was established (report of the second plenary session). This group presented its report at the fourth plenary session in Sundsvall in

1990, and recommended several actions that the IPCC might wish to undertake in order to increase participation by developing countries in IPCC activities. The most important measure towards this end was the provision of funds to support developing-country participation financially.

With the restructuring in 1992, it was agreed that instead of a Special Committee on developing-country participation "the special situation of the developing countries should ... be given attention as part and parcel of all the work carried out by the Panel and its groups (working groups/subgroups/task forces)" (IPCC Task Force on IPCC structure, third session, Doc.2, 1992, item 4.3.3). It was especially emphasised that developing-country participation in the bodies constituting the scientific core of the IPCC should be increased. In the discussion at the second session of the IPCC task force on IPCC structure, the point was made that "while the scientific-technical nature of the IPCC work should be preserved, to the extent possible, by all available means, this may pose some difficulties for the developing countries since they normally do not have experts in every area of IPCC interest; hence, in the case of developing countries, general competence should be emphasized over specific expertise" (draft report of the second session of the IPCC task force on IPCC structure, agenda item 2.4).

In 1992, developing-country participation was integrated at all levels of the IPCC, and the Special Committee on Developing Countries was dissolved. This organisational change implied, first, a restructured and enlarged bureau in order to achieve better geographical balance. The bureau would now consist of the chairman and the two vice-chairmen of the panel, the co-chairmen and vice-chairmen of the three IPCC WGs, and six regional representatives (Africa, Asia, Europe, North and Central America, South America and the Southwest Pacific) (report of the eighth plenary session, item 6.12.2). Second, the organisational change implied an increased emphasis on the recruitment principle of geopolitical representativeness in all IPCC bodies at all levels. It was, for instance, decided that the co-chairmen and vice-chairmen of the WGs "should be chosen so that the developing and developed countries are equally represented", however, with "due regard to the need to maintain the scientific-technical integrity of the IPCC" (IPCC task force on IPCC structure, third session, Doc. 2, item 5.13). Furthermore, in 1993 it was specified that "insofar as practicable", the composition of the group of lead authors "should include at least one expert from a developing country" ("IPCC procedures for preparation, review ...", item 2).

Judging by the number of developing-country delegates to the IPCC plenary sessions, the efforts to increase developing-country participation seem to be relatively successful. Developing-country participation has increased substantially since 1988: While only 14 non-OECD participants attended the first plenary session in 1988, 48 attended the fourth plenary session in 1990, and 98 non-OECD participants (of a total of 215 participants) attended the eleventh plenary session in 1995 (Agrawala, 1998). The largest expenditure of the IPCC is the financial support of developing-country participation. In 1996, for instance, the cost of a three day IPCC meeting was estimated at 750,000 Swiss Francs, of which 630,000 were budgeted for the financial support of 100 participants from developing and transitional-economy countries ("Proposal for an IPCC budget for 1996-1997", Doc. 9, eleventh plenary session). While participation of developing-country experts in the lead author teams has increased substantially since the first IPCC assessment, the aim of including one developing-country expert in all lead author teams where WGI is concerned has not been successful. The geographic representativeness of the IPCC is still, however, considered to be high.

It should be noted, however, that an apparent representativeness – measured in terms of participation in IPCC meetings – may be an insufficient measure of the inclusion of developing countries in the generation of a problem diagnosis. That is, as argued by Joyeeta Gupta, the source of an imbalance lies in a structural imbalance in the generation of knowledge because of lack of domestic science in the South (1997: 164). If there is an imbalance in state-of-the-art knowledge, in the sense that it generally reflects knowledge that is generated in the North (because the North generally produces more scientific publications), this imbalance will be replicated in the knowledge base and the problem diagnosis generated by the IPCC.

6.4.4 Mechanisms for Conflict Resolution

The provision of mechanisms for the resolution of conflicts that arise during the process of science–policy interaction is particularly important in relation to the development of a consensual problem diagnosis. Mechanisms for handling two main types of conflict situations are in particular demand: first, conflicts that may arise between scientists, especially over the extent to which the knowledge base provided within the policy-making context can be considered a fair representation of state-of-the-art knowledge; and second, conflicts that may arise between policymakers and between scientists and policymakers over how this knowledge base should be communicated.

In chapter 4 we suggested that mechanisms for conflict resolution in processes of science–policy interaction are established to the extent that: a) arenas for an interactive dialogue between scientists and policymakers are established; b) deliberations can be transferred to informal arenas; and c) policymakers acknowledge scientists' expertise as a basis for an "authoritative" decision-making procedure in disputes related to the interpretation and presentation of scientific knowledge.

In the IPCC process both of these types of conflicts have been experienced; there have been disagreements and differences of opinion among scientists over the substantive content of the assessments, and disagreements and conflicts between policymakers and between policymakers and scientists over the appropriate formulations by which to communicate the findings in the assessments (note that only WGI is analysed here). Since 1992, the explicit elaboration of the policy implications of the knowledge base has not been part of the IPCC's agenda. Thus, there have been few conflicts directly related to this aspect. It is clear, however, that possible policy implications in many cases have been the source of the disagreements and conflicts experienced. This is above all illustrated in the conflict between WGs I and II over the use of paleoclimatic analogues as a forecasting technique, which will be discussed in more detail below.

In the IPCC process the instruments suggested in chapter 4 for handling these situations – the establishment of arenas for an interactive dialogue, the development of informal arenas, and decision-making procedures based on informal positions of authority – have been employed with relative success for handling conflicts over modes of presentation between scientists and policymakers. These instruments seem less applicable, however, for handling situations of disagreement and conflict between scientists over the development of a knowledge base that constitutes a fair representation of state-of-the-art knowledge within the relevant fields. Other mechanisms, designed to resolve disagreement and develop consensus within the scientific community in general, have been employed in the IPCC for handling situations of this kind (see also chapter 7, section 7.2.2).

Within the IPCC, the WG plenary is the main arena where conflicts between policymakers and between scientists and policymakers surface. As pointed out above, the WG plenary serves its most important function as an arena for an interactive dialogue between scientists and policymakers. In general, the WG plenary represents an opportunity for scientists and policymakers to discuss and develop agreement on how a piece of information is best presented and communicated. Thus, the main function of the WG plenary is as a forum where scientists and policymakers can meet regularly and where important aspects of the knowledge base on climate change are developed through an interactive dialogue between scientists and policymakers.

In some cases, however, the differences of opinion expressed in these discussions reflect (conflicts of) interest in the policy area. In such cases the discussions at WG plenaries may take the form of straightforward negotiations. This is primarily indicated by the extent of political strategic behaviour demonstrated by many delegations. We have already discussed how especially the large oil-producing countries have behaved strategically in accordance with their national interests in this policy area in WGI plenaries. Their persistent efforts to amend the reports towards an increased emphasis on GHGs other than CO_2 and aspects of uncertainty seems to qualify as strategic behaviour in this regard. On several occasions these delegations have, for instance, made attempts to bargain the formulations of the report by announcing their acceptance of one formulation on the condition that their concerns regarding another formulation are accommodated (see also chapter 7, section 7.2.4). There are, however, also other examples of strategic behaviour in WGI. Especially at the fifth plenary session of WGI in Madrid, the USA made several attempts to strengthen the formulations of the SPM, sometimes quite radically (recall the proposal to change the sensitivity range, see footnote 5, section 6.3.2). While these attempts may primarily be considered an effort to counterbalance the proposals made by the oil-producing countries, it is also likely that they may have been motivated by a concern linked to the role of the IPCC reports as a basis for US domestic climate-change policy. The Clinton Administration has been engaged in a continuous battle with the Republican majority of the Senate over domestic climate-change policy. Strong and unambiguous statements from the IPCC are necessary for the development of a proactive US climate policy. Their behaviour in the WGI plenary may, therefore, have been partly motivated by this concern and may as such represent an example of a "double-edged diplomacy" strategy (see Evans et al., 1993). This was indirectly confirmed by a US delegate at this WGI plenary session, although he strongly emphasised that their proposals were scientifically substantiated and that a true reflection of scientific knowledge in the SPM was their primary concern (personal communication with US delegate at the meeting).

The presence of strategic thinking and behaviour in WGI plenaries taken into account, the establishment of informal arenas and smaller side meetings is important. It should be noted, however, that a transfer of the discussions to informal arenas is a very important tool for being able to get through the agenda of plenary meetings at all, even in situations characterised by little or no controversy. A transfer of the discussions to informal arenas, therefore, represents an important tool in the development of science–policy consensus on the assessment summaries in the IPCC in general, which also may be utilised as a mechanism for the resolution of conflicts if and when they occur. As noted in our discussion above, the establishment of informal arenas is a central feature of the proceedings of the IPCC's WGI. At the WGI plenary meeting in Maastricht in 1994, this procedure worked very well, and the informal discussions that took place provided scientists and policymakers with an opportunity also to discuss questions related to the location of the boundaries between science and politics (see section 6.4.2; Shackley and Skodvin, 1995). Paradoxically, however, this mechanism is difficult to employ in situations where the level of conflict is high. As we have seen, at the fifth plenary session of WGI, which was a conflict-filled and difficult meeting, this procedure was not accepted by some delegations.

Finally, a decision-making procedure based on positions of authority is also a central characteristic of the proceedings of WGI. As discussed above, lead authors usually get the last word in situations of disagreement between scientists and policymakers over the formulations of the summaries. The lead authors' scientific authority is generally accepted by policymakers as a basis for this decision-making procedure, and if policymakers wish, they can record their dissenting view in a footnote. So far, no delegation has chosen to do so in WGI (see also section 6.3.2).

Thus, depending on their nature and intensity, conflicts are resolved in WGI plenaries in one of three ways: the subject is discussed, negotiated and agreed upon in plenary; the discussion is transferred to an informal arena in smaller side meetings; or, if neither of these procedures work, the issue is decided upon (vetoed) by the lead authors with an opportunity for delegations to record their dissenting views.

These instruments are primarily designed for resolving disputes and conflicts between delegations and between delegations and scientists. They are not suitable for handling scientific disputes. In WGI, disputes between scientists have been handled by methods corresponding to the methods by which the scientific community usually deals with scientific dispute, although perhaps with a stronger determination to find compromises and formulations capable of integrating all scientifically based viewpoints.

The most important mechanism in this regard is the review procedure. Peer review is the main mechanism for scientific quality control in all scientific endeavour. The review procedure constitutes the backbone of the IPCC process and is essential for the scientific credibility and authority, as well as the political acceptability, of IPCC reports. The IPCC has developed an *extended*, two-tiered, review procedure; the peer (expert) review, and the government review. The extended review procedure serves a dual function. First, the peer review is the main mechanism for ensuring that the assessments constitute a fair representation of state-of-the-art knowledge within relevant fields and disciplines, and is acknowledged as such by the scientific community. Thus, the peer review is the main link between IPCC scientists (lead and contributing authors) and the scientific community at large. While the main function of the peer review is scientific quality control, it may also be regarded as a mechanism at least for visualising points of dispute within the scientific community, which are then subsequently discussed and resolved at lead- and contributing-author meetings. Points of dispute are usually resolved either by developing generous intervals of uncertainty or by attempting to develop compromises that reflect the major (scientifically substantiated) viewpoints of the community of scientists (see section 6.3.2). Second, the government review serves to bring the assessments and their summaries under the adversarial scrutiny of parties with conflicting interests. This is of particular importance with regard to the summaries, since they represent a *choice* of which conclusions (of the assessments) the community of scientists finds it important to communicate to policymakers (recall that the summaries are a 10–20 page summary of a 3–400 page assessment), and thus represent an *interpretation* of the knowledge base and a first step towards the transformation of this knowledge into decision premises in the form of a comprehensive problem diagnosis. The government review thus provides governments with a first opportunity to comment on the interpretation of the knowledge base as represented in the summaries, and provides the teams of lead authors and the leadership of the WGs with an opportunity to intercept conflicts before the reports reach WG plenaries by incorporating comments beforehand.

The review procedure serves an important role for the IPCC's capacity to handle disputes and conflict both within the scientific community and in the science–policy debates that take place in WG plenaries. Thus, while the review procedure itself cannot be regarded as a mechanism for handling and resolving conflicts within the IPCC, it serves an important role in bringing to light points of dispute and potential conflict which then can be brought up and handled in the appropriate forums and in appropriate manners depending the nature of the disagreement.

The IPCC review procedure has several shortcomings, however. One has been referred to as the problem of the "silent majority" (Agrawala, 1998b). The nature of a review implies that reviewers are primarily asked to communicate points of criticism. Reviewers are therefore unlikely to communicate explicitly their support of the various parts of the drafts that they do not comment upon. By incorporating all comments brought forward in the review, the leadership of the WG may make changes in the texts in response to just a few critical remarks. This may, therefore, generate new points of dispute at subsequent meetings (lead- and contributing-author meetings or plenary meetings). On the other hand, by incorporating comments selectively, reviewers may feel that their work was to no avail, that they have been used as "alibis" by the IPCC[15] and that they have no real influence potential. In the report of the eighth plenary session it is for

[15] Very often the objectivity and impartiality of the IPCC process is documented with reference to the number of contributors and reviewers of the reports.

instance reported that "some reviewers in the past had felt that their comments had not been considered adequately and hence objected to being listed as this implied concurrence" (agenda item 5.2). This may therefore also generate new conflict at subsequent meetings. This problem is not easily resolved either, since a system where reviewers are asked to communicate both criticism and support would be unmanageable in practice. The employment of external agents to supervise the incorporation of comments could perhaps modify the problem, however, at least in the sense that external judgement on the weight of the comments brought forward is introduced in the process (Agrawala, 1998b). Another problem associated with the IPCC review procedures is that (the same) reviewers may communicate their comments through several channels (Agrawala, 1998b). Governments often delegate review work to their (national) scientists. Since these scientists also may be approached by the IPCC as expert reviewers they get the chance to communicate their comments (at least) twice. Since scientists from industrialised countries dominate the scientific core (at least in WGI) (both as lead, contributing and reviewing scientists), this deficiency may lead to a preponderance of comments from industrialised countries. This problem may be relatively easily resolved, however, by developing a system that restricts the options for reviewers to comment only once.

In sum, then, the procedures adopted by the IPCC in the development of assessments do represent institutional arrangements that may be utilised as mechanisms for dealing with points of dispute and conflict both between scientists and between scientists and policymakers. The main mechanism for recognising and bringing to light dispute within the community of scientists is the peer review procedure. The government review on the other hand, serves to bring the assessments and, most importantly, their summaries under the adversarial scrutiny of parties with conflicting interests. This represents the first step towards acquiring policymakers' acceptance of the assessments and their summaries, subsequently completed at WG plenary meetings. At WG plenaries the establishment of informal arenas and the employment of a decision-making procedure based on positions of authority where lead authors have a (near) veto are utilised as instruments for generating consensus. It is important to note, therefore, that the procedures by which disputes are handled serve different functions in the assessment process. While the handling of disputes within the scientific community largely is related to the development of the assessment and to ensuring that it constitutes a fair representation of state-of-the-art knowledge within relevant fields, the handling of disputes between scientists and policymakers is largely related to the interpretation of the knowledge base and hence the transformation of the knowledge base into a comprehensive problem diagnosis.

It should be noted, moreover, that these procedures cannot be regarded as mechanisms for conflict resolution per se. The primary function of the peer review, for instance, is scientific quality control. The sheer breadth of this review (with literally hundreds of scientists involved) implies, however, that the normal review traditions of scientific work in general are not applicable, and also indicates the additional function of the review related to the handling and resolution of conflicts within the scientific community. Moreover, while strategic behaviour is very much present at WGI plenary sessions, we have not found any examples of outright bargaining and horse-trading over the summary texts corresponding to political negotiations. Hence, mechanisms for resolving dispute within the IPCC seem to be more subtle and much less explicit than in negotiations proper, although they are certainly represented in the proceedings of the IPCC.

The subtlety of conflict resolution within the IPCC taken into account, it may be useful to illuminate the handling of conflicts in an explicit example. Below, therefore, a more detailed account of the conflict between WGs I and II over the use of paleoclimatic analogues as a forecasting technique, which took place in the development of the first IPCC assessment, is presented.

The Particular Case of the Conflict over Paleoclimatic Analogues

In 1989–90 a conflict arose between WGs I and II in the preparation of the first IPCC assessment over the use of past warm climate intervals as an analogue of a future greenhouse modified world. This conflict is interesting in several respects: It is an example of how possible policy implications and concerns could infuse the scientific debate, and how a conflict with an obvious political dimension was handled by scientists and the leadership of (especially) WGI and the panel. It also serves to illustrate the views of some central actors in the IPCC process on the role of the IPCC in terms of providing policy advice and how the IPCC reports would be of most use to policymakers. This conflict is, therefore, discussed in some detail.[16]

The conflict arose after a WGII meeting in Moscow early in 1989 where the question of paleoclimate analogues as a forecasting technique dominated the meeting. The technique was particularly advocated by the Russian (paleo-) climatologist Mikael Budyko, whose reconstruction of a world with $2xCO_2$ atmospheric concentrations, based on paleoclimatic data, indicated far more favourable climate conditions than those derived from General Circulation Models (GCMs). In particular, Budyko's "paleoclimatic scenarios", which indicated increased precipitation in general and hence

[16] The discussion is based on internal correspondence and other informal material obtained and copied with permission from the Technical Support Unit of WGI in Bracknell in September, 1994.

more favourable conditions for agriculture, were suggested to imply an increased wheat productivity in the (then) Soviet Union by up to 50% (assuming CO_2 concentrations of 425 ppm) (internal correspondence; *New Scientist*, 26 August, 1989). On the basis of these results, Budyko allegedly argued that reducing the burning of fossil fuels would be "a crime against humanity" and that "the most advantageous decision would be not to retard but to intensify the growth of CO_2 concentrations which would rapidly result in the reconstruction of the Pliocene climate optimum conditions which are more favorable than the modern ones" (internal correspondence within the IPCC). Budyko, supported by the Russian Chairman of WGII Jurij Izrael, insisted that these paleoclimatic scenarios be used by WGII as a basis for their impact studies. Moreover, Izrael and Budyko criticised WGI for not including a discussion on the paleo-analogue forecasting technique in their report. A WGII meeting in Nalchic in March 1990 was, for instance, asked to endorse a statement of dissatisfaction with the WGI report in this regard (internal correspondence).

The reliability of this technique and the validity of these results in the context of a human-induced climate change in modern times were strongly criticised and questioned by other scientists. *New Scientist* reports for instance that "American climatologists are extremely sceptical about Budyko's predictions. They warn that data about past climates are extremely fragmentary. Data for Budyko's maps of Pliocene conditions were collected from remains differing in age from 3 million to 4–5 million years old" (26 August 1989). The main criticism was that the climatic impact of heating by two entirely different mechanisms cannot be assumed to be the same. While temperature variations in prehistoric times were caused by slight fluctuations in the orientation of the Earth's axis of rotation with respect to the sun coupled with subtle changes in the shape of the orbit around the sun, current temperature increases would be caused by a human-induced increase in the concentrations of GHGs in the atmosphere (internal correspondence).

The conflict also had a dimension of high politics, since the question of paleo-analogues as a forecasting technique for future climate conditions was also a subject of the US–USSR joint Report on Future Climates under the US–USSR Environmental Agreement of 1987 to be finalised early 1990, and which was to be submitted to the IPCC for their use in preparing the first IPCC assessment. Thus, the question of paleo-analogues became a sensitive issue. In an internal letter within WGI dated May 1990, the dilemma of American scientists involved both in the US-USSR report and the WGII report is summarised: "If they are too dismissive of paleo-analogs they run the risk of sabotaging the whole US–USSR cooperation. On the other hand they recognise that, if they leave the door open on paleo-analogs as predictions, they will undermine the conclusions of the WGI report." In the same letter, the leadership of WGI emphasised the importance of ensuring "that the conclusions of their [the US–USSR] report were consistent with ours; we do not want Izrael to get up at the final IPCC plenary and wave the US–USSR report around as proof that they were right all along." To the leadership of WGI, it became important to resolve the question within WGI, and, most of all, to keep the conflict out of the IPCC plenary.

In a confidential letter dated 19 June 1989, a WGII lead author voices his concern over the situation on the paleo-analogues, after results indicating that a 50% increase in wheat productivity could be expected in the Soviet Union in a future warmer world had been distributed at a WGII meeting: "If this material is included in WGII's report, there are serious implications for policy (especially for the "do nothing" strategy) ... This issue is getting very messy; and it is shot through with politics. I *would be unwilling* to include estimates of impact on agriculture under paleoclimatic scenarios *unless* there were an unequivocal scientific case made that such scenarios offer a valid means of characterising the future" (emphasis in original, internal correspondence between WGs I and II).

In November 1989, an extraordinary two-day WGI workshop, in its entirety devoted to the issue of paleo-analogues, was arranged. Approximately 20 scientists participated, including Budyko. In the invitation letter to the meeting it is pointed out that, "you will be aware that the use of paleo-analog techniques as climate forecasts is strongly supported by Russian scientists, but has been criticised by others. This is not just an academic issue; the forecasts from the two techniques are different and may have different policy implications, and so it is important that the issue is resolved amongst those who are best equipped to deal with it" (invitation letter to the Bath meeting in November 1989 from Dr. John Houghton, Chairman of WGI).

The Bath meeting was sceptical towards the reliability and validity of paleoclimatic analogues as a forecasting technique for the climatic conditions to be expected in a future greenhouse-modified world, and the conclusions of the meeting were communicated to the chairman of WGII: "Firstly, the extent of the database in all but the Holocene period is very limited. Secondly, due to the differences in the forcing and boundary conditions, the Holocene, Eemian and Pliocene periods cannot be easily used as detailed analogs for future climate in the next century. Given these conclusions, we are therefore unable to advocate the use of analogs from the past as indicators of the spatial changes in temperature and precipitation to be expected in the future" (telex from Geoff Jenkins, co-ordinator of WGI to J. Izrael, of 4 December 1989).

The issue was subsequently also discussed at a lead author meeting of WGI in Brisbane in December 1989, and it was agreed that a discussion on the paleo-analogue technique and the validity of predictions based upon it as compared to GCM integrations should be included in a sub-section of the WGI report: "Because their results are quite different, it seems likely to me that the conclusion will favour one or the other of the two techniques. It would be of no use to policymakers if we offered them two quite distinct predictions without guidance on which we believed was most credible" (draft letter from the chairman of WGI to the chairman of WGII). This concern over the usefulness of their assessment to policymakers was also touched upon by Dr. Houghton in his welcoming address to the Bath meeting: "We must bear in mind that the IPCC report will provide direct and urgently needed advice to policymakers, and we must therefore be as clear as possible in our conclusions" (informal conference papers in preparation for the Bath meeting).

Having agreed that a sub-section of the WGI report should deal with the paleo-analogue technique, the discussion then concerned who should author this section. In this regard the chairman of WGI makes it clear that WGI did not consider Mikael Budyko a candidate for this task: "Professor Budyko has made what are seen to be some very extreme statements about policy toward climate change, based on the paleo models; suggesting that to cut back on fossil fuel burning would be a crime against humanity, for example. I am concerned, therefore, that someone who has taken such a committed and public stance may not be appropriate to author an assessment which has to be based on disinterested scientific judgement if it is to be accepted" (draft letter from the chairman of WGI to the chairman of WGII).

The conflict did not end with the decision to include a discussion of the paleo-analogue method in the WGI report, however. In a telex to WGI of 21 December 1989, the chairman of WGII points out the following: that all possible methods should be used for climate predictions; that the IPCC (plenary) decided this and that only they can change it; that the Bath meeting was not long enough to discuss the paleo-analogue method properly; that the negative conclusions of WGI on the paleo-analogue method coupled with the restrictions in the credibility of the GCMs made the task of the IPCC impossible; and finally, that the issue should be brought up at an IPCC plenary meeting.

After this turn of events, panel chairman Bert Bolin initiated the establishment of an informal working group including Budyko, Bolin himself and two other WGII scientists who also had been involved in the US-USSR project (personal communication with Bert Bolin). During their discussion it soon became clear, according to Bolin, that Budyko operated under instructions from Izrael, and that he could not comment or make decisions on several of the involved aspects without first consulting with Izrael (personal communication). This was, according to Bolin, the end of the story. The Russians' insistence that this method be employed as a basis for the impact studies of WGII were mostly politically motivated, and revealing this, they no longer had a case.

The paleo-analogue method is discussed in section 3.5.1 (Cubash and Cess, 1990: 83–84) and 5.5.3 (Mitchell et. al., 1990: 158–159) of the first WGI Assessment Report. It is concluded that "the paleo-analogue approach is unable to give reliable estimates of the equilibrium climatic effect of increases in greenhouse gases as suitable analogues are not available, and it is not possible to allow for the deficiencies in the analogues which are available. Nevertheless, information on past climates will provide useful data against which to test the performance of climate models when run with appropriate forcing and boundary conditions" (Mitchell et. al., 1990: 159).

6.5 Enhanced Effectiveness?

The time has come to confront the crucial question of the analysis: Have the institutional arrangements of the IPCC contributed to enhancing the effectiveness of the science–policy dialogue? To what extent has the institutional apparatus of the IPCC served the four main functions, and has it had the desired effect?

6.5.1 Score

The above discussion indicates that the institutional apparatus of the IPCC as a whole to a large degree has served to: maintain the scientific autonomy/integrity of scientists, ensure a large degree of involvement between scientific and political actors, ensure the geographic representativeness of the institution, and to some extent provide mechanisms for the resolution of conflicts. Hence, the institutional apparatus of the IPCC seems to function well in terms of the four dimensions identified in chapter 4 as determinants of the capacity of institutional arrangements to enhance the effectiveness of the science–policy dialogue. The above discussion strongly suggests, however, that within the IPCC these various functions are served by different bodies, i.e., at different decision-making levels within the institution. The IPCC constitutes a multi-tiered system with a high level of formal differentiation between roles and functions – notably the actual provision of a knowledge base and the transformation of this knowledge base into a comprehensive problem definition and diagnosis, agreed upon by both scientists and policymakers, which may subsequently serve as a premise for policy decisions. One major conclusion of this analysis, therefore, is that the institutional design of the IPCC provides for an important differentiation between roles and functions at different decision-making levels which enables the institutional apparatus to serve a set of seemingly incompatible functions. This explains, in particular, the difficult combination achieved within the IPCC of both separating and integrating science and politics.

The actual provision of IPCC assessments takes place within the scientific core of the WGs. At least as far as WGI is concerned, the scientific core is well protected against purely politically motivated influence. Participants at this decision-making level are largely scientists active in research, funded through the regular channels for scientific funding at the national level. Above all, the scientific core of WGI enjoys a high level of both formal and operational autonomy, with the norms and standards of scientific endeavour in general as an influential guide and with no systematic governmental control.[17] On the other hand, there are few formal channels of communication between scientists operating within the scientific core of WGI and the more policy-dominated decision-making levels of the institution, apart, that is, from the role of lead authors in WGI plenaries. Participants within the scientific core of the IPCC are largely "bench scientists" with minimal direct contact with policymakers and the policy dominated bodies within the IPCC. In this regard, the scientific core of WGI is characterised by a low level of policy involvement and science–policy interaction. Overall, therefore, the scientific core is characterised by a high level of both formal and operational autonomy, while it is characterised by a low level of involvement and science–policy interaction. The converse case characterises the situation in the WGI plenary. Participants in WGI plenaries are largely government officials and scientists with governmental affiliations, selected largely on the basis of administrative position, whose participation is funded by national governments, operating under a relatively strong governmental control. These aspects are all clear indications of a low level of scientific autonomy (modified by the central role of lead authors at this decision-making level). On the other hand, there are strong formal and informal links between the WGI plenary and policymakers, not least as a function of participation and recruitment mechanisms at this decision-making level. This indicates a relatively high level of involvement.

The functions between these decision-making levels are also differentiated. While the scientific core has its primary function in the provision of the actual assessment, the WGI plenary is mainly concerned with what could be regarded as the *interpretation* of this assessment, as reflected in the choice of emphases and formulations in its *summaries*. Taking this into account, the procedure by which this interpretation is

[17] I have not investigated more subtle modes of governmental control over research activities at the national level, such as, for instance, funding mechanisms which discriminate between research projects on the basis of purely political and non-scientific considerations.

developed, moreover, with intense science–policy and policy-internal discussions that sometimes have a clear aspect of strategic behaviour and thus strongly resemble straightforward negotiations, the function of the WGI plenary may be seen as linked to the *transformation* of the knowledge base into decision-making premises, rather than the provision of the knowledge base itself. While the WGI plenary has little direct influence on the provision of the actual Assessment Report, both scientists, as represented by lead authors, and policymakers, as participants in WGI plenary sessions, have, as we have seen above, a significant potential influence over the assessment summaries, which are developed through an interactive dialogue between lead authors and policymakers in WGI plenary meetings.

Through this differentiation of roles and functions between decision-making levels, a combination of scientific autonomy and science–policy integration is achieved *within* the WG. After the 1992 restructuring, this balance between scientific autonomy and science–policy integration can also be assumed to characterise the other WGs of the panel. With this institutional set-up, the WG plenaries serve an important buffer function in terms of separating science and politics within the WGs of the panel.

Before 1992, a balance between scientific autonomy and policy involvement was also achieved horizontally, across WGs, particularly between WGI and the old WGIII. While WGI managed to maintain the scientific autonomy of its scientific core and restrict policymakers' influence on the substantive content of the knowledge base, the old WGIII became politicised at all decision-making levels. However, as discussed above, the old WGIII served an important function as a "policy arena" during a phase when no policy arena was formally established. In this regard the old WGIII provided an important link between scientists and policymakers during a phase when it was not at all obvious how this link could otherwise have been provided without compromising the scientific autonomy of the other WGs. Moreover, there seems to be a general understanding, especially within the leadership of WGI and the leadership of the panel, that the old WGIII also served an important function as a "lightning rod", redirecting policymakers' attention and influence attempts away from the other WGs (personal communication). During this phase, the IPCC largely operated without formal rules and procedures and the process had not yet developed its own dynamics. The buffer function served by the old WGIII may therefore have played a vital role for the ability of WGI to maintain the scientific autonomy and integrity of its scientific core during a phase when the process was particularly vulnerable to political "contamination".

6.5.2 Effect

This analysis suggests, therefore, that the institutional dimensions of autonomy and involvement serve different functions in science–policy interaction processes. While autonomy primarily seems to facilitate the development and *provision* of a knowledge base, involvement seems to facilitate the *transformation* of the knowledge base into inputs to processes of policy-making. Moreover, this analysis suggests that these dimensions may not be as hard to combine as initially assumed. The institutional dimensions of autonomy and involvement may be combined to the extent that institutional arrangements can be differentiated according to function (see also Andresen et. al., in press). This is precisely what has been accomplished in the case of the IPCC.

One major aspect of a functional differentiation is the establishment of institutional buffers that serve to separate the scientific and the more policy-oriented functions of the institution. In the IPCC context (after 1992), the WG plenary seems to serve as a buffer between the (most) politicised decision-making level of the panel, the panel plenary, and the scientific cores of the WGs. Before the 1992 restructuring, the old WGIII served a different, but equally important, buffer function which may have increased the ability of the other WGs to maintain the scientific autonomy of their scientific cores. As pointed out by Miles, the establishment of institutional buffers between research results and their utilisation for regulation and distribution of costs is important, particularly from the perspective of science itself (Miles, 1989). This analysis suggests, however, that a functional differentiation also may be useful from the point of view of policymakers (see also Underdal, in Andresen et al., in press).

The institutional set-up of the IPCC, and particularly the capacity of institutional arrangements to balance and combine scientific autonomy and science–policy integration, seems to have contributed substantially to the extent to which policymakers have acknowledged the scientific authority of the knowledge base, and accepted its substantive conclusions as factually valid. In this case, policymakers' confidence in the research results communicated by scientists seems to be drawn from at least two main sources: first, the scholarly competence, integrity and independence of the scientists involved in the process; and second, the adversarial scrutiny by actors and parties representing conflicting interests in the policy area. In an area as conflict prone as climate-change policies, the latter mechanism seems at least as crucial as the first.

In the IPCC, adversarial scrutiny is accomplished through different procedures. First, the extended and open review procedures adopted by the IPCC serve an important function in this regard, in terms of ensuring that the reports and their summaries are subjected to the scrutiny of the scientific community at large *as well as* ensuring the adversarial scrutiny by governments representing potentially conflicting interests at a relatively early stage in the assessment process (before the reports are submitted and discussed in plenary sessions). Second, the assessment summaries are subjected to an intense scientific and political scrutiny in WG plenary sessions. The function of WG plenaries as an arena for an interactive science–policy dialogue is the most visible mechanism for ensuring adversarial scrutiny by parties with conflicting interests. Third, adversarial scrutiny is ensured through the recruitment and participation procedures adopted by the IPCC, particularly with regard to the institutional arrangements developed over time to ensure the geographic representativeness of the panel. The aspect of geographic representativeness also serves to illuminate the differentiation of institutional arrangements according to function in the sense that a geographic balance within the scientific cores of the WGs is sought to the extent that it is compatible with the requirement of scholarly competence. The fact that the scientific core of WGI does not entirely reflect the geographic composition of the parties in the policy area does not seem to have had any significant impact on the acceptability of the knowledge base to policymakers. A similar imbalance in the WGI plenary, on the other hand, which, in contrast to the scientific core, is primarily concerned with the transformation of the knowledge base into an applicable input to the policy-making process, would most probably have had a significant (negative) impact on policymakers' acceptance of the knowledge base.

Thus, the subjecting of the Assessment Reports and particularly their summaries to adversarial scrutiny by parties with conflicting interests is a major feature of the institutional design of the IPCC.

This emphasis on the provision of critical review procedures and mechanisms for adversarial scrutiny within the institution seems also to require a system for handling the conflicts and points of dispute that can be assumed to arise in its aftermath. This analysis of the IPCC, however, finds that the institution largely relies on implicit, rather than explicit, mechanisms for conflict resolution. The mechanisms for handling conflicts and dispute within the IPCC are implicit in the sense that procedures are designed and integrated into the needs of the assessment process, but that these procedures also can be utilised for purposes of conflict resolution if and when conflicts occur. The conflict between WGs I and II in 1989–90 demonstrates that the institutional arrangements of the panel, and particularly the individuals occupying central roles within the institution, have been capable of handling relatively difficult conflicts arising in the interface of science and politics. Mechanisms for handling disputes and conflicts thus seem to be embedded in the institutional apparatus designed for the development and provision of assessments. The capacity of the IPCC to tap this instrumental potential, however, seems to a large extent to depend upon the behaviour of individual agents within the system. While the institutional design of the panel to some extent provides institutional instruments whereby conflicts may be handled and resolved, their actual utilisation seems to depend upon the initiative and diplomatic competence of individuals and occupants of central (leadership) roles (the aspect of leadership is further discussed in chapter 7).

The ability of the IPCC to submit its reports to the adversarial scrutiny of parties with conflicting interests in the policy area is also a function of the high level of political involvement characterising the IPCC process. In the IPCC, "involvement" largely refers to policymakers' involvement in the scientific process, not scientists' involvement in the policy-making process. In the public debate of climate-change science and policies, the IPCC process has maintained its image as a scientific body and a scientific process. This is slightly paradoxical given the extensive governmental and political participation characterising the process in general, and, especially, the most public and visible parts of the process, namely the WG and panel plenaries towards which most public attention is drawn. Policymakers' degree of involvement in the IPCC process is, above all, indicated by the fact that it is the panel plenary – the most political decision-making level of the institution – which decides the rules of procedure of the panel.

The climate change issue touches upon vital national interests and thus involves a high potential for serious political conflicts. Furthermore, the issue is scientifically complex and associated with profound scientific uncertainty. Scientific findings are to a large extent open to interpretations, with a corresponding risk, from the point of view of policymakers, of information distortion. The combination of these features seems to have implied a desire by governments to ensure scientific objectivity and disinterestedness, while at the same time, maintain a certain political control over the process. This is achieved by submitting the products of the process to the adversarial scrutiny of a broad set of actors, and by submitting the functioning of the process itself to a political control through the rules of procedure.

A key factor in this regard is the intergovernmental status of the IPCC, which permits governmental participation at the main decision-making levels. Through this institutional device, policymakers are permitted to participate in the development of the knowledge base. They thus have access to very detailed information, not only concerning the nature of the problem at hand, but also concerning the manner in which this knowledge has been brought about. Especially the latter aspect constitutes very important information in a negotiation setting characterised by conflict because negotiating parties are then provided a basis for evaluating the reliability of the information they receive regarding the nature and diagnosis of the problem at hand. In this case, moreover, the intergovernmental status of the IPCC implies that policymakers have significant influence on the manner in which the knowledge base is provided – the rules and procedures according to which the IPCC operates. They can therefore contribute to the establishment of rules and procedures that, in their view, serve to increase the reliability of the knowledge base. This, however, can also be used as an instrument in efforts to undermine the IPCC's credibility. That is, delegations deprived of scientifically based arguments to support their political position can accuse the IPCC of violations of their own rules of procedure in the development of the knowledge base and hence demand that conclusions and findings should be deleted. There are several examples in the IPCC's history that illustrate that delegations, especially from the largest oil-producing countries such as Saudi Arabia and Kuwait, have adopted this strategy. The political control over rules of procedure has, therefore, been utilised by some governments as a tool for delaying the process, and has posed a difficult challenge for the IPCC leadership (this point is further elaborated in chapter 7, section 7.2.4).

This design could have been devastating for the development of a scientific knowledge base on global warming because it left the process particularly vulnerable to undue political influence on the substantive content of the knowledge base. When such an influence cannot be traced, the scientific authority and the (potentially) central role of lead authors as exemplified in WGI plenaries seems to be an important explanatory factor. Moreover, the ability of the leadership of the IPCC to retain control over the scientific process and serve as an authoritative guide in the incremental development of the rules of procedure seems to be another important explanatory factor (see chapter 7). Succeeding in this, moreover, the institutional design of the IPCC serves the very important function of providing an arena for a broad science–policy dialogue. Especially the WGI plenary serves this function. WGI plenary meetings do not only constitute an opportunity for sceptics and opponents to employ delaying tactics and attempt to modify aspects of the reports that do not correspond to their political positions, it also serves as a forum where scientists and policymakers can discuss various interpretations of the findings, seek clarification where necessary, exchange opinions concerning potential policy implications and discuss which future tasks the IPCC should embark upon. In this regard, the institutional design of the IPCC seems to constitute a crucial factor in the explanation of the high level of acceptance among policymakers of the scientific knowledge base provided by the IPCC. While it is methodologically impossible to control this effect, it seems very unlikely that the knowledge base could have gained the level of acceptance that it actually has in this case if the knowledge base had been provided, for instance, by an independent body in complete isolation from the political process on climate change.

This proposition is supported by the leadership of the IPCC in personal communication with this author. The intergovernmental status of the IPCC is emphasised by the leadership as its most important institutional device. Without it, it is maintained, policymakers wouldn't have had to listen and the potential for influencing policymakers would have been significantly reduced (personal communication). This point has also been made by Bob Watson, chairman of WGII until 1997 and the new chairman of the IPCC as of September 1997:

> "I believe the IPCC process is much, much more powerful than the single-agency approach. The most important thing when you have an assessment process is that it has to be credible to all stakeholders. They may not all agree with the outcome, but if they're all part of designing the process in the beginning, they'll be more willing to let the chips fall where they may." (Bob Watson in interview with *Science*, September, 1997)

The level of involvement of governments in the provision of a knowledge base on climate change has been subject to continuous debate. In an interview with *Nature,* Bill Hare, climate policy adviser for Greenpeace, states that he is not against government involvement but feels that "at present, the balance of influence is too much in their favour" (*Nature*, vol. 381, June 1996). In the same interview, Sir John Houghton, Chairman of WGI, on the other hand, reportedly stated that "any move to reduce political involvement in the IPCC would weaken the panel and deprive it of its political clout". He is quoted as saying, "The presence of government scientists is vital to the IPCC. ... They own the findings. If governments were not involved, then the documents would be treated like any old scientific report. They would end up on the shelf or in the waste bin" (*Nature*, vol. 381, June 1996).

This analysis thus suggests that the institutional design of the IPCC process constitutes one important explanatory factor to the outcome in this case. Important aspects in this regard are the extent to which institutional arrangements are differentiated according to the main function of the body – thereby serving to combine scientific autonomy and science–policy integration – and the adversarial scrutiny of the knowledge base by parties representing conflicting interests in its transformation into a comprehensive problem definition and diagnosis. Thus, it is also argued here that the institutional arrangements, and their capacity to serve the four main functions identified in chapter 4, have served to enhance the effectiveness of the science–policy dialogue of the IPCC process.

A crucial condition for this effect, however, is the leadership functions served by occupants of key roles in the process. The leadership provided by individual actors is, in a sense, the "glue" of the system. Indeed, it seems that the explanatory power of institutional design for the outcome of this process is entirely dependent upon the capacity of individual actors to provide leadership, both in the development of the assessment, its transformation into decision premises for policy decisions and in boundary roles between the scientific and the policy dominated decision-making levels of the institution. This proposition is the subject of further analysis in the next chapter.

Chapter 7

Agent: Leadership Performance in the Intergovernmental Panel on Climate Change

7.1 Introduction

The capacity of the institutional arrangements of the IPCC process to enhance the effectiveness of the science–policy dialogue seems to a large extent to be dependent upon the capabilities of individual agents to serve leadership functions. This chapter will investigate this proposition further.

Chapter 4 suggests three theoretically derived functional areas in which appropriate leadership can be assumed to be in particular demand:

- the development of the knowledge base;
- the transformation of the knowledge base into premises for policy decisions;
- the provision of boundary-role leadership in the interface of science and politics.

Three main issues are addressed in this chapter: First, the extent to which these leadership functions have been provided in the IPCC process and what these leadership roles seem to have implied in terms of behaviour is addressed in section 7.2. Second, the extent to which the performance of leadership is linked to/conditioned by specific institutional arrangements is addressed in section 7.3. Finally, section 7.4 investigates the impact the performance of leadership has had for the functioning of the IPCC system and particularly the extent to which leadership performances have served to enhance the effectiveness of the science–policy dialogue.

7.2 Identifying Leadership Performance

The identification of leadership performance in complex social processes and the investigation of causal relationships between this mode of behaviour and process outcomes is perhaps one of the most methodologically and practically challenging research tasks in political science. Since the phenomenon subject to study is a mode of behaviour whose occurrence usually cannot be traced and identified through written documentation, the study of leadership is a highly data-intensive research task requiring intimate knowledge of the process under investigation, preferably by personal observation and/or access to informants with this kind of insight who also are willing to share their insight in this regard. In this particular research area, moreover, the lack of guidance provided by other systematic theoretical and empirical investigations further complicates the task. The role and impact of various forms of leadership behaviour has to some extent been theoretically explored with regard to processes of negotiation (Young, 1991; Underdal, 1991a,b; 1994; Malnes, 1995; see also chapter 4, section 4.5.1). No studies in which this question has been systematically pursued in the field of science–policy interaction, however, were uncovered in the course of this research. Moreover, the theoretical explorations of leadership behaviour in negotiation analysis have only to a slight extent been supplemented and "tested" in empirical analyses (but see, *inter alia*, Fermann, 1997; Skodvin Hegdal, 1992). For these reasons, the aim of the following analysis is modest.

The provision of leadership functions requires an agent. As pointed out by Young, "leaders are individuals, and it is the behavior of these individuals which we must explore to evaluate the role of leadership" (Young, 1991: 287). The first critical task, therefore, is the identification of candidates for leadership behaviour which thus merit closer investigation. In chapter 4, leadership was defined as "an asymmetrical relationship of influence in which one actor guides or directs the behavior of others toward a certain goal over a certain period of time" (Underdal, 1994: 178). The actual provision of leadership seems to be a function of at least two main determinants: capabilities and behaviour (skill). The essence of leadership is to convert one's capabilities or leadership potential into actual influence. Regardless of the capabilities of the actor, the task requires skill. Conversely, no matter how skilful, behaviour alone cannot outweigh lack of capabilities. In this section, therefore, the investigation of leadership is approached in two consecutive steps. First, the question of which capabilities that seem to be required to perform the leadership functions suggested in chapter 4 is explored. Since it is practically impossible to investigate all actors that might hold this potential, this analysis seeks to identify which actors operating within the IPCC system are most likely to hold this potential. Thus, the analysis is focused on individuals operating in formal roles which provide them with a particular potential to serve leadership functions in the IPCC process. It should be emphasised that this is a discriminatory tool that must be employed in order to keep the analysis manageable. Second, the investigation focuses on these (groups of) individuals to identify whether they have served a leadership role as suggested, and, if so, what this leadership function seems to imply in terms of explicit behaviour. In this part of the analysis also, however, a research strategy that is compatible with the manageability of the task has been adopted. Even after the search has been restricted to individuals that are likely to hold the relevant capabilities, the group of potential candidates is, in some cases, too large to conduct a detailed investigation of the behaviour of each and every one. Also, the data are simply too limited to perform a detailed study of the behaviour of each candidate. Therefore, this analysis focuses on which types of behaviour by occupants of formal roles in the process seem to be associated with the provision of leadership functions. This method allows investigation of the extent to which the leadership functions suggested above have been provided by actors operating in formal roles within the IPCC system, although it may not be possible to point exactly to which individuals that have actually served the leadership functions. Moreover, the question is addressed by both "positive" and "negative" evidence. In some cases there are clear indications of how specific types of behaviour have contributed to the provision of leadership functions and how this has affected the process and its outcome. In other cases, however, the question is explored by concentrating the investigation on cases where the process obviously has not functioned very well, and thereby investigating and demonstrating the likelihood of leadership behaviour in cases where it has. In this manner the analysis identifies the functions for which a leadership is likely to have been provided and also suggests the explicit modes of behaviour the various leadership functions seem to have implied.

7.2.1 Actor Capabilities

Chapter 4 identifies three main categories of functions for which the provision of leadership may be assumed to be in demand: (i) leadership functions associated with the development of the knowledge base; (ii) leadership functions associated with the transformation of the knowledge base into decision-making premises; and (iii) leadership functions that may contribute to link these two stages of the process together, which may be defined as boundary-role leadership in the interface of science and politics. Leadership functions directed at the provision of mechanisms for conflict resolution are considered an integral part of all categories, but are assumed to be particularly associated with categories (i) and (ii).

These three categories of leadership functions constitute the "demand" side. The question addressed in this section is which actor capabilities are associated with the "supply" side of the equation? Which actor capabilities seem to be required for individual agents to be capable of providing these leadership functions, and which actors within the IPCC system hold these capabilities?

In general, three types of capabilities seem to be required for the provision of these leadership functions: (i) substantive knowledge; (ii) both personal and formal capabilities for procedural management; and (iii) interpersonal and human skills both to co-operate and to induce co-operation among participants in the process. While their emphasis may vary according to the function in question, all of these capabilities seem to be required to perform a leadership role successfully.

More specifically, the capabilities required to perform a leadership function in the provision of the knowledge base seem to be mainly linked to substantive knowledge in the form of formal scientific qualifications. The development of a knowledge base in which state-of-the-art knowledge within relevant fields and disciplines is adequately represented requires the in-depth knowledge of relevant research and research findings only scientific education and experience can provide. Thus, this leadership role can be assumed to be provided only by scientists active in research in these areas.

Of the actors operating within the IPCC system, contributors and lead authors as well as WG chairs and panel chairs are likely to hold this capability. Furthermore, within this group of likely candidates, the lead authors of WGI are the actors most likely to hold this capability. Thus for all practical purposes, the search may be restricted to this group. It is therefore suggested that the lead authors of WGI are the actors within the IPCC system who are most likely to hold a potential for providing leadership in the development of the knowledge base.

The capabilities required to perform a leadership role in the transformation of the knowledge base into premises for policy decisions also seem to be linked to scientific qualifications. A leadership role in this category, however, can also be assumed to require an ability to communicate across the boundaries of science and politics. Thus, the role seems to require substantive knowledge not only in scientific terms, but also in terms of a general knowledge and familiarity with politics as a system of behaviour as well as more specific knowledge about the political dimensions associated with the particular issues subject to discussion. In this regard, actors associated with both of these systems of behaviour are likely candidates for the provision of leadership functions in this category.

In the IPCC process, the transformation of the knowledge base into premises for policy decisions largely takes place in the WG and panel plenaries. As a first cut, therefore, those actors who do not participate at these decision-making levels may be excluded. The residual group of candidates comprises delegates, lead authors, WG chairs and the panel chair. Since it is not practically possible to investigate the scientific and political background of all delegates at these decision-making levels, this group is excluded from further search, although there may be candidates within this group who actually hold a potential to perform a leadership role within this category. Moreover, while lead authors, as was discussed in chapter 6, serve a very central role in WG plenaries, very little is known about each lead authors' capabilities in terms of their association to politics as a system of behaviour, for performing a leadership role in this category. The WG and panel chairs, on the other hand, are most likely to hold these capabilities. Thus, the analysis will concentrate on WG chairs and the panel chair, although, given the central role of lead authors at WG plenaries, the door will also be held open to possible leadership functions provided by this group.

The capabilities required to perform leadership functions in boundary roles in the interface of science and politics can be assumed to be very similar to those required to perform leadership functions in the second category, i.e., a professional association to both science and politics as systems of behaviour. However, this category of leadership functions is also associated with access to specific arenas, and hence formal capabilities for procedural management. In particular, within the IPCC, this category of leadership can only be provided by actors with access to all the decision-making levels of the system, from the scientific core to the panel plenary. In addition, at least with regard to a leadership function directed at the development of institutional arrangements, access to the IPCC bureau is also required, since it is this body in which institutional and procedural issues are first addressed. Moreover, a leadership function directed at the provision of communicative links between scientific and political bodies also requires access to the main political bodies external to the IPCC system, the INC and COP in particular. Given this set of necessary capabilities to perform a leadership function in this category, there is only one group of likely candidates left: WG and panel chairs. Moreover, with regard to the provision of leadership functions that require access to the INC/COP, the analysis may be focused almost entirely on the panel chair, although WG chairmen on some occasions have assisted the panel chair in his reporting to these bodies.

The capabilities required to perform a leadership function directed at conflict resolution in the development of a knowledge base and its transformation into decision premises cannot be specified on the basis of formal requirements. Rather, this leadership function may be assumed to a larger extent to be associated with an individual's interpersonal and human skills in co-operating, inducing co-operation among participants and handling and resolving conflicts. It may therefore be assumed that the capacity of lead authors and WG and panel chairs to provide leadership functions in these categories is also dependent upon their (personal) capabilities to induce co-operation and resolve the disputes and conflicts that arise in the course of the process. This leadership function, therefore, is considered an integral part of the leadership functions associated particularly with the first and second categories.

The formal roles to which the provision of leadership functions may be assumed to be associated have now been identified: Lead authors in WGI are the most likely candidates for the provision of leadership functions associated with the development of a knowledge base; WG and panel chairs are the most likely candidates for the provision of leadership functions associated with the transformation of the knowledge base into decision premises, although lead authors also may play a leadership role in this regard; and WG and panel chairs are the most likely candidates also for the provision of boundary-role leadership in the interface of decision-making levels within the IPCC system, and the panel chair is the most likely candidate for the provision of boundary-role leadership in the interface of the IPCC and political bodies external to the IPCC system. The next section will investigate more closely the extent to which these actors have succeeded in converting this potential to serve leadership functions into the provision of actual leadership.

7.2.2 Leadership in the Development of the Knowledge Base

Within the scientific core of WGI lead authors play a central role in the provision of the assessment reports. They are key players in the selection of contributors and expert reviewers (and also, on some occasions, in the selection of other lead authors) and, above all, they bear the main responsibility for incorporating into the assessments all scientifically substantiated viewpoints and findings of the scientific community, as communicated to them by contributors and reviewers, in a representative and balanced manner. In the 1993 rules of procedure it is, for instance, pointed out that

> the essence of the lead authors' task is synthesis of material drawn from the peer reviewed literature, generated at workshops or submitted by contributors. Lead authors may not necessarily write original text themselves, but they must have the proven technical ability to develop

text that is scientifically sound and that faithfully represents, to the extent that this is possible, contributions by a wide variety of experts.

It is further emphasised that, "principles governing IPCC work require LAs to record views which cannot be reconciled with a consensus view but which are nonetheless scientifically or technically valid" ("Tasks and responsibilities for lead authors, ..." cited in the report of the ninth plenary session, Appendix G, Annex 2, item 1). Thus, lead authors bear a main responsibility for ensuring that the knowledge from relevant fields is reflected in the knowledge base in a manner that is representative of the state-of-the-art. In performing this task, lead authors exercise a substantial amount of *judgement*, and both the scientific and the political community's appraisal of the knowledge base hinges on their perception of how well equipped the lead authors are in making this judgement.

The capabilities required for lead authors to perform a leadership role in the assessment process is mainly linked to their formal scientific qualifications. In this regard their potential to provide leadership is a function of the recruitment mechanisms employed in the selection of lead authors to the assessment reports.

While lead authors are selected from lists of nominations by governments, the actual choice lies with the scientific leadership of the WGs. Scientists not on the nomination list are never chosen as lead authors, but the IPCC leadership have on some occasions approached governments to have particular scientists nominated (personal communication with Bert Bolin). The procedure whereby lead authors are chosen has become increasingly formalised during the course of the process, but even with the formalisation of procedures in 1993, there are relatively few formal requirements guiding the choice. In the 1993 rules of procedure, the qualifying criteria for the choice of lead authors were specified. It emphasised that due consideration be given to scientists "known through their publication or work". The "technical ability" of the lead author and their "ability to work to deadlines" are also emphasised as important criteria. Finally, it is pointed out that teams of lead authors "should reflect a fair balance of different points of view". The leadership of WGI emphasise four main criteria in their choice of lead authors (personal communication): the scientific merits of the candidate; his/her ability to write English; his/her ability to work to deadlines; and that he/she is not too dominating. With regard to the latter, the WGI leadership emphasise that the lead author should be able to listen to others, not only write his/her own research into the chapters, and that the texts written by lead authors must reflect the discussions of the lead- and contributing-author meetings. Of these four main criteria, the scientific merits of the candidate, judged in terms of the traditional criteria whereby scientific work is evaluated by the scientific community at large, such as publications with scientific quality control and citations, is pointed out as the most important. These criteria are to a large extent also emphasised as the most important by lead authors themselves (personal communication). In addition, one lead author emphasised that in choosing lead authors they do not only look at specific qualifications, they also look for a team of lead authors that work well together: "We need one who's a good leader, we need one who's good at writing, we need one who knows everything, and we need one who's a hard working scientist and knows all the details and references" (personal communication).

WGI has to a large extent succeeded in recruiting distinguished and highly respected scientists with an unquestionable scientific authority among their peers as lead authors to their assessments. Most of the lead authors in WGI have impressive CVs and publication lists. Their highly praised scientific qualifications imply that most lead authors in WGI have the capabilities required to succeed in a leadership role. Their ability to convert this potential into actual leadership, however, seems to depend on the extent to which they enjoy the trust of the scientific community in their role as lead authors to the IPCC assessments. This, in turn, seems to be linked to their personal capabilities and skills in terms of incorporating scientific findings from a broad field of research including areas to which they are not themselves experts and, particularly, in terms of handling scientific dispute and controversy that may arise in the course of this process. Lead authors do not, therefore, serve a leadership role by virtue of their scientific qualifications alone; their personal qualifications linked to the process of generating the assessment and their ability to acquire the scientific community's trust that all scientifically substantiated viewpoints will be reflected in the assessment seems to be equally important.

For a scientist to perform well in his role as lead author, therefore, a set of scientific and personal qualifications are required. On the scientific side, he should, ideally, be a respected scientist within his own area of expertise and possess a broad knowledge and overview of the field at large. On the personal side, he should be able to listen to others and, in a sense, include scientific arguments and viewpoints with which he does not necessarily agree. On the practical side, moreover, he must be able to prioritise IPCC work for (sometimes long) periods of time. A lead author's capability of combining these qualifications determines his/her success in performing a leadership function in the development of the assessment.

The importance of these characteristics is exemplified by the difficult situation that arose in the aftermath of the shift of lead author in WGI in 1990. In February 1990, the lead author (LA1) of one of the sections of the First Assessment Report of WGI resigned from his position. The actual circumstances around his resignation are not well known, but apparently his resignation was linked to a dispute over the inclusion of some new material in the section. At the time of LA1's resignation, the assessment process had been going on for one year. The lead author (LA2) who replaced LA1, had not participated in the numerous lead- and contributing-author meetings and other activities that had taken place during this time. He was therefore placed in the unenviable position of finalising an assessment process in which he had not taken part. He went about this task by re-editing a final draft. In this re-editing, which apparently implied a major revision of the section, new material was added and old paragraphs were rewritten in such a way that the discussions of contributing authors during the previous year were not well reflected. This aroused serious concern among contributing authors to the section. In a letter to the co-ordinator of WGI, dated 2 April 1990, one contributor (C1) writes:

> There are a few new issues [LA2] is trying to insert into the Section ... (including a brand new conclusion for the executive summary – I suggest it be deleted). Some of this has been discussed thoroughly in our previous meetings. I have basically reiterated my views in my comments, but since this whole process is now going down the wire, there is no way I can be sure that [LA2], or anyone else, will properly take my comments or those of the other contributors, into account as the report rushes toward publication. I contemplated asking that my name be removed from the Section entirely for this reason. Instead, I decided to invest more of my time and come up with detailed comments in the hope that the section will come out as a reflection of what we agreed upon in [the lead author meeting] and not some new thing composed at the last minute. ... This Section and the report are too important for last-minute activity to be determining the content after *one full year* of meetings and drafts and discussions. (emphasis in original).

C1 is careful to emphasise, however, that it is not the scientific qualifications of the new lead author he questions, but his ability to draft a text that properly reflects the previous discussions in the group of scientists in which the new lead author has not participated: "I want to make it clear that this is in no way a reflection on [LA2] or the daunting task that has been dumped in his lap. His reputation speaks for itself and a scientist of his calibre could make a clear and important contribution to this effort". In a letter of the same date to the new lead author concerning the incorporation of the new material, C1 points out that: "These statements were negotiated between various contributors and represent what they wanted to say about [this matter] in terms of their model simulations. As such, they stand as written and must be consistent with new material added". After conversations with other contributors to the section (here referred to as C2 and C3), and after an unsuccessful effort to arrange a meeting with the new lead author of the section, C1 expresses his concern again to the co-ordinator of WGI in a letter dated 18 April 1990:

> This has turned into a rush to judgement and it is unclear to those of us who contributed to [this section] how it will actually come out. Clearly this is not a good situation ... [C2] was perhaps even more agitated about [this section] than [C3] and I are, and he also feels that the IPCC 'process' has broken down to such a degree now that maybe we all should remove our names from the Section since material and conclusions will be represented in ways that may call into question the integrity of the Section and how it was written. The fact that none of us has had a chance to discuss any of this in person with the new lead author means that, in a worst case scenario (and one that hopefully would not happen), [this section] would turn into an editorial view ... representing one person's interpretation and may in no way be indicative of an assessment carefully formulated by the research community.

Apparently this unfortunate situation was resolved by redistributing the final draft of the Section to contributing authors for comments, and also by supplementing the lead authorship of the section with two additional lead authors who were familiar with the previous discussions of the group of contributing scientists. Though some new material was added by LA2, the material already written for the section at the time of the lead author shift remained the main bulk of the chapter (personal e-mail correspondence with contributor). The contributing authors who expressed their concern about the new lead author's mode of operation and who threatened to withdraw from the process, are cited as contributing authors in the published report. One of the concerned contributing authors expresses (in personal e-mail correspondence) that he was satisfied with the final product.

This incident clearly illustrates that scientific brilliance is not enough to be a good lead author in WGI. The personal capacity and skill of the lead author to provide leadership in the development of the assessment in terms of developing text which appropriately incorporates the viewpoints and discussions of the contributors to their satisfaction is equally important. This is a question of both procedural management – in terms of developing procedures appropriate for providing contributors with opportunities to make their viewpoints known and for generating contributors' trust that their viewpoints will be taken into account and reflected in the reports – and interpersonal and human skills for dealing with disputes and for inducing co-operation among contributors. It is, moreover, important to note that the turbulence of this section does not seem typical for the IPCC WGI assessment process – neither in the preparation of the First Assessment nor at later stages in the IPCC's history. In general, the assessment process within the scientific core of WGI has run much more smoothly than in this case. In general, therefore, the appointed lead authors of WGI seem to have been well qualified, both in terms of scientific and personal capabilities, to serve the leadership function required for the process to be effective. This incident nevertheless demonstrates that the leadership role served by lead authors in the assessment process was not simply a function of recruitment procedures, and hence institutional arrangements. There is a substantial element of personal skill in a good lead authorship, and the large extent to which this personal element seems to have been provided also seems to have contributed substantially to the functioning of the assessment process within the scientific core of WGI.

7.2.3 Leadership in the Transformation of Scientific Knowledge

Within the IPCC, the transformation of knowledge into premises for policy decisions is mainly associated with the adoption of the Summaries for Policymakers (SPMs) in WG plenaries. At this stage of the process, three potential leadership functions may be envisaged: First, facilitating and guiding the discussions and negotiations themselves may be in demand, since the discussions in WG plenaries are, as was demonstrated in the preceding analysis, difficult encounters between scientists (lead authors) and government officials where a host of hidden agendas may affect the course (and effectiveness) of the debate. Second, for the same reasons, the resolution of conflicts and disputes that arise in the actual drafting of texts may also be in demand at this stage. Finally, "knowledge brokering" in the more explicit form suggested by Litfin – behaviour more directly associated to the development of decision-making tools – may also be explored at this decision-making level.

As we have seen in our previous analysis, the possible policy implications that may follow from the findings, conclusions and formulations of the SPMs are an ever-present consideration made by participating delegations in WG plenary discussions. While the policy implications may be obvious in some cases, policymakers usually do not have a full overview over which policy implications that may follow from the reports of the IPCC, or even in which policy area implications may follow. Policy-makers thus seem to engage in these discussions with a very broad perspective of the policy areas for which the reports may have implications. A minor change in the drafted texts, maybe even without any scientific significance, can spawn a discussion totally out of proportion, judged in relation to the apparent significance (or insignificance) of the proposed change, because of hidden agendas (personal observation). While such discussions normally have little impact on the actual texts, they can be extremely time consuming. At WG plenaries, the WG chair, assisted by the panel chair, play an important and difficult role in terms of preventing the discussions from trailing off into dead ends. In this regard, WG plenary discussions require a chair with a concentrated and clear mind and an awareness of the ever-present political dimensions of the issues being discussed.

This was particularly demonstrated at the WGII plenary session in Nairobi in 1994, where the proposal of a minor change in the drafted text resulted in a substantial delay in the proceedings due to the prolonged discussion that followed. At this meeting, the Executive Summary and the SPM of the "IPCC Technical Guidelines for Assessing Climate Change Impacts and Adaptations", prepared by WGII to the 1994 Special Report (Carter et al., 1994), were submitted for approval by the WGII plenary. It is important to emphasise that the report in question was an assessment of methodologies for carrying out impact and adaptation studies; it was not an assessment of impact and adaptation studies per se. The discussion concerned whether or not the definition of "adaptation" (to climate change) in paragraph 8.1 should refer only to "adverse" effects (as in the draft), or should refer both to "adverse" and "positive" effects. Although many delegations argued that it was ridiculous to talk about adaptation to positive effects, the WG chair agreed that the methodologies used to carry out impact studies are, in principle, neutral with regard to the nature of the effect. After some discussion, it was therefore agreed by the meeting that in the definition of "adaptation" adopted in the report, both adverse and positive effects should be mentioned (paragraph 8.1 in the Executive Summary of the report).

Towards the end of the meeting, however, the issue was reopened in the context of a discussion of paragraph 8.2.5 on methodologies for quantifying the performance of adaptation measures.[1] Saudi Arabia and Russia argued, as had been agreed upon with regard to paragraph 8.1, that this paragraph should also refer both to adverse and positive effects. The lead authors tried to abort the discussion by emphasising that while methodologies in principle are neutral towards the nature of the effect, it would hardly be necessary to develop a methodology on how to quantify the performance of adaptation measures towards positive effects. Russia, however, pursued the issue with a long intervention on the necessity of performing risk assessments also on

[1] The last sentences of the paragraph were discussed in particular: "Uncertainty analysis and risk assessment are also considered at this stage. This step is a prelude to developing strategies, each of which promoted one or two key objectives, while attempting to minimize the adverse effects associated with the remaining objectives" (paragraph 8.2.5 in the Executive Summary to the draft report).

positive effects. So far, the discussion had concentrated on paragraph 8.2.5. At this point, however, Brazil, which had opposed the reference to positive effects in paragraph 8.1 in the previous discussion, now intervened and also reopened this issue. The WG chair expressed concern about reopening a closed issue, but he did ask the plenary whether there was any support for the Brazilian proposal. That was undoubtedly a mistake. A long discussion over paragraph 8.1 followed, and the meeting gradually polarised into two "camps": one for and one against the reference to "positive" effects. Moreover, during the course of the discussion it also became apparent that not all delegations realised that the report was actually a recommendation on methodologies, not policies, and that it was a recommendation to technicians, not politicians. Several efforts by both the WG chair, the panel chair and the lead authors to clarify and explain this difference were unsuccessful in terms of ending the discussion. Thus, the debate was entirely beside the point, it had no scientific significance whatsoever, and it was also totally "out of control". Finally, the WG chair proposed to establish an informal working group to sort out the issue of paragraph 8.1. This proposal was met by the following statement from the Saudi Arabian delegate (who wanted the reference to positive effects in 8.1 to remain in the document): "OK. We have now adopted the practice of reopening paragraphs that already have been adopted. Fine. Tomorrow we will reopen the discussions on the points from 1–8. This practice should also be transferred to the other WGs, and tomorrow we are having a meeting in WGIII" (cited from personal notes from the meeting). This situation was resolved by the panel chairman's decision that two alternative statements were to be developed and included in the minutes of the meeting (not the actual report), with a record of how many delegations supported each statement. In the actual report, paragraph 8.2.5 was redrafted so that no reference to the nature of effects was made, while paragraph 8.1 remained as agreed upon earlier.

There are two aspects of this discussion that merit particular attention. First, the discussion is to a large extent incomprehensible in a scientific perspective (at least its duration and intensity). It was inflamed by at least two different politically motivated hidden agendas: First, the Saudi Arabian delegation's insistence that positive effects should also be specifically mentioned[2] is easy to comprehend in the perspective of their position in the policy area. Wherever possible, Saudi Arabian delegates advocate the inclusion of statements that emphasise uncertainty, GHGs other than CO_2, and, as in this case, the potential positive effects resulting from a human-

[2] It should be noted that Saudi Arabia did not stand alone on this position, but in the course of the discussion, it increasingly became a debate between Saudi Arabia on the one side and Brazil on the other, with the WG chair, the panel chair and the lead authors desperately trying to sort it out.

induced climate change. Brazil's position – that of opposing the inclusion of a reference to positive effects in relation to adaptation measures – is comprehensible from a logical point of view. Brazil's part in the debate, however, and its complete lack of willingness to compromise, also seems to have been nurtured by a politically motivated hidden agenda. Apparently, the Brazilian delegation was concerned about the possible implications of the language chosen in the IPCC reports for the basis upon which countries could apply for GEF-funds to finance adaptation measures. They were concerned that with an explicit reference to positive effects in the IPCC report, the door would be open to GEF-applications for financial support of adaptation measures to positive effects. Apparently, therefore, they were concerned about increased "competition" over GEF funds for adaptation measures (personal communication with delegates to the meeting, not Brazilian delegates). The Brazilian delegation which, after all, had "logic" on their side in the debate, also suffered, however, through their continual miscomprehension of the distinction between methodologies and actual impact and adaptation assessments. This miscomprehension may also have served to prolong their participation in the discussion.

Second, it is very likely that this difficult situation could have been avoided entirely with firmer leadership from the WG chairman. The WG chairman made two main mistakes: First, he overlooked a proposal from the Netherlands early in the discussion over paragraph 8.1, repeated again early in the discussion over 8.2.5, to delete all reference to the nature of the effect, and simply refer to "effects". With regard to paragraph 8.2.5, this was actually the final outcome. It is likely that had this proposal been pursued by the WG chairman at an early stage of the discussion, the matter could have been resolved relatively easily and the entire discussion over paragraph 8.2.5 could have been avoided altogether. Second he permitted the meeting to re-open the question of paragraph 8.1, when raised by Brazil, after this discussion had been closed and resolved in agreement a few hours earlier. That was the spark that fired the discussion.

This discussion is interesting in the sense of being a typical example of how discussions in the WG and panel plenaries of the IPCC can trail off into time-consuming and intensely political debates without scientific significance. This continual risk constitutes a particular challenge for the WG and panel leadership in chairing the discussions. Their ability to provide firm leadership in this regard seems important for the course of the discussion. This leadership role seems to be linked to at least three abilities: First, the WG and panel chairs must recognise a good compromise when they see one, preferably at an early stage of the discussion. Second, their ability to judge whether a compromise is needed at all, and whether a proposal actually constitutes a "good" compromise, is linked to their awareness of and ability to pre-empt the political dimensions of the issue under discussion. Finally, this leadership role is also a question of something as trivial as "perfect timing" in "clubbing" the issues on the agenda through (the WG chair knocks on the table with a little hammer when an issue has been agreed upon). The delegations must be given sufficient time to consider the scientific merit of a proposal, but not so much time that they can ponder the breadth of possible policy implications over a wide scope of policy areas. In general, therefore, the leadership role takes the form of guiding the discussions in the right direction, and preventing them from heading into blind alleys and hidden agendas.

While there are numerous examples of similar episodes in the IPCC's history, the WG and panel chairs generally seem to have managed to provide this mode of procedural management. Most importantly, when such discussions do take place, they have managed to prevent them from turning into complete deadlock situations. Somehow they have always found a way out of the (political) tangles such discussions may cause. During the course of the process they have also acquired a certain skill in foreseeing such situations and pre-empting the discussions before they manifest themselves as issues of principle and major policy concern in the minds of delegates. Finally, they have learned to tackle such situations if and when they occur. In general, therefore, the leadership provided by WG and panel chairs in terms of providing guidance in the discussions at these decision-making levels and, particularly, their skill in finding solutions and preventing such discussions from turning into deadlock situations seems to have been important for the functioning and outcome of the IPCC process.

While the WG and panel chairs seem to have served an important function in terms of keeping the discussions of WG plenaries on a "scientific track", the role of lead authors seems to be equally important in the actual drafting of text. The operating procedure at WG plenaries implies that lead authors have a substantial amount of influence over the SPMs. In this regard, the rules of procedure place a main responsibility on the lead authors to develop texts whereby the concerns of delegations are accommodated to the greatest extent possible without compromising the scientific justification of the text. The analysis in chapter 6 is evidence of the substantial amount of time and effort lead authors have invested in this task. For instance, when controversies over appropriate formulations arise between lead authors and delegations in the WGI plenary, the discussion is usually moved out of the formal context of the plenary meeting into informal "drafting groups" in which lead authors and delegations negotiate the paragraphs in question until they can come up with a formulation upon which all can agree. If they do not succeed, and a lead author "rules" against a proposed change, delegations have the opportunity to record their dissenting view in a footnote to the text.

The little extent to which delegations have made use of this opportunity may be regarded as an indication of a relatively well functioning procedure to which the capability of lead authors to accommodate different concerns in the drafting of text seems to have contributed substantially. Had the lead authors simply "vetoed" the texts of the SPMs, delegations would most probably have employed their right to record their dissent much more frequently and the political acceptability of the texts would have decreased accordingly. On the other hand, had the lead authors simply inserted the proposed changes by governments, the scientific legitimacy of the text would be compromised and the scientific authority of the IPCC reports would be reduced accordingly. Thus, the balance achieved through the leadership function of lead authors in the drafting of texts seems to be essential.

This point is useful also to illustrate the nature of the negotiations that take place within the IPCC process and, especially, the nature of the "compromises" that are developed. The analysis of both this and the preceding chapter demonstrates that negotiations have taken place in the IPCC process. Unlike distributive negotiations, however, where two or more parties give and take to find middle-ground solutions between otherwise incompatible positions, negotiations within the IPCC concern the scientific accuracy of proposed text – specifically the scope of validity of scientific findings and the precision with which to express them. Thus, disputes are resolved largely by integrating different concerns and by enhancing the precision with which findings are expressed, particularly in terms of their scope of validity, rather than by developing compromises in the traditional sense where parties win some and lose some in solutions found between opposing positions. In this process of integration and qualification of scientific findings, the guidance provided by lead authors, particularly on the basis of their scientific capabilities, seems to be essential.

Thus, WG chairs, panel chair and lead authors each have a significant potential for providing different forms of leadership in the process whereby WG plenaries approve the SPMs. Their ability to realise this potential seems to have served to facilitate the discussions in WG plenaries. In this manner, these actors also seem to have been instrumental in "brokering" knowledge in the sense of facilitating its transformation into premises for policymaking. There is no apparent evidence, however, of "knowledge brokering" in the more explicit form suggested by Litfin in her studies of the international regime on ozone depletion.

In Litfin's study, "knowledge brokering" refers to the ability of individuals to frame and interpret scientific information so that its applicability to policymakers is increased (Litfin, 1994). In particular, she finds that in the ozone process, this activity, typically conducted by individuals who operate at low or middle levels of governments or international organisations, is directed towards the development of decision-making tools. The prime example is the shift of focus which occurred around 1986 from calculations of the "ozone depleting potential" (ODPs) of various gases and substances to the "chlorine loading" methodology. This shift of focus implied that scientists were able to assess at which levels of chlorine concentrations damage to the ozone layer had occurred, and accordingly, at which levels the ozone layer would recover. In the development of this methodology, EPA scientist John Hoffman was particularly instrumental.

The methodology had significant policy implications – primarily in the sense that the scenario modelling field was opened up "to anyone with a powerful desktop computer" (Parson, 1991: 27). In this regard, the chlorine-loading methodology represented a decision-making tool to a much larger extent than calculations of ODPs, for which policymakers were dependent upon major input from scientists. Perhaps even more importantly, the shift of focus had significant policy implications in terms of the change in perceptions of risk and caution that followed. Employing the chlorine-loading methodology, EPA scientist, John Hoffman, presented in 1986 a study that showed that an 85% cut in current emissions would be required simply to freeze atmospheric chlorine levels (Litfin, 1994: 100). When the issue was framed in these terms, and especially with the detection of the ozone hole in 1985, Litfin finds that "suddenly, a phaseout did not seem like a drastic proposal" (Litfin, 1994: 100). A similar example of the development of scientific methods that may serve as tools for policymaking, may also be found in the development of the "critical loads" methodology in the international regime on long-range transboundary air pollution (LRTAP) (see for instance, Wettestad, in Andresen et. al., in press).

In the IPCC process, there are no similar examples of knowledge brokering. The climate change process is still in the "ODP"-phase, in the sense that the major policymaking tool developed by scientists so far, is the Global Warming Potential (GWP) index. The GWP index was introduced as a tool for policymakers to compare the potential of the various well-mixed source gases to affect climate. While the GWP method is a simple and understandable formula by which to transform emissions of various gases into one measure, it is still associated with ambiguity and scientific debate (Shackley and Wynne, 1997). The GWP method cannot be employed for all species (notably those whose lifetime is shorter than the time for mixing in the troposphere) and the crucial choice of time horizon in the employment of GWPs, which is largely left to policymakers, may have substantial policy implications[3] (Skodvin and Fuglestvedt, 1997). The GWP method did,

[3] If one common time horizon is to be chosen in the use of GWPs, regulations on either long-lived or short-lived gases will be favoured by the choice. For instance, the GWP for

however, serve as the scientific basis for the comprehensive approach adopted in the Kyoto protocol (with a 100-year time horizon), which implies that parties may chose to implement their commitments by regulating emissions of any of six specified (groups of) greenhouse gases.[4] At this stage of the policymaking process on climate change, however, the ambiguities associated with the GWP method may have been as important for its implementation in the Kyoto protocol as its *virtues* as a policymaking tool (see Shackley and Wynne, 1997).

Moreover, no individual agent can be identified as particularly influential in the development of the GWP index, at least not within the framework of the IPCC process. On the other hand, as demonstrated in Litfin's study, knowledge brokering mainly took place at arenas outside the formal framework of the science–policy dialogue on ozone depletion (Litfin, 1994; see also Skodvin in Andresen et. al., in press). Hence, this mode of leadership might not be expected within the framework of the IPCC and may have been provided at arenas beyond the scope of this analysis.

7.2.4 Boundary-role leadership in the Development of Institutional Arrangements

In general, the demand for input from research to processes of policymaking provides the scientific community with incentives for co-ordinating research activities and establishing some sort of mechanism to make research hypotheses and findings applicable, available and useful to policymakers. Conversely, the supply of relevant and applicable research findings and analyses seems to stimulate the demand for research input. At this intersection, there seems to be considerable scope for boundary-role leadership directed towards organising supply, stimulating demand and matching demand and supply (Underdal, in Andresen et. al., in press). In the

---

methane (CH_4), whose atmospheric lifetime is approximately 12 years, is 56 for a 20 year time horizon while it is 6.5 for a time horizon of 500 years. In contrast, the GWP for perfluoromethane (CF_4), whose atmospheric lifetime is approximately 50,000 years, is 4,400 for a time horizon of 20 years while it is 10,000 for a time horizon of 500 years (Schimel et. al., 1995: 121). If we assumed that countries had uniform emission profiles – in the sense that their emissions mainly consist of either long-lived or short-lived gases – negotiations on time horizons could acquire the character of zero-sum games: what the one wins, the other loses. It is more realistic to assume that countries would want to choose time horizons that permit mixed measures, directed towards both long-lived and short-lived gases. It is difficult, however, to see how this can be achieved in practical terms since different time horizons might compromise the comparability of different gases, which would be the essential point of the whole exercise (see Skodvin and Fuglestvedt, 1997; Shackley and Wynne, 1997).

[4] Carbon dioxide (CO_2), methane (CH_4), nitrous oxide (N_2O), hydrofluorocarbons (HFCs), perfluorocarbons (PFCs) and sulphur hexafluoride (SF_6).

context of the IPCC process, the provision of boundary-role leadership seems to be particularly important in an institutional system that serves both to separate and, to some extent, integrate science and politics. Two functions may be envisaged for this role: First, to design institutional arrangements that achieve both a separation and an integration of science and politics. Second, to provide communicative links between the various decision-making levels of the system as well as communicative links between the IPCC and the political community as represented by the INC/COP.

As discussed in chapter 6, the institutional arrangements comprising the IPCC system are differentiated according to function. The IPCC is organised in three decision-making levels that serve different functions in the assessment process: While the provision of the knowledge base mainly takes place within the scientific core of the institution, the transformation of this knowledge base is mainly linked to the proceedings of the WG and panel plenaries. This differentiation of institutional arrangements based on the function of each decision-making level in the assessment process seems to have had significant implications for the outcome of the IPCC process. In particular, the capacity of the institutional apparatus of the IPCC to both separate and integrate science and politics, achieved through this differentiation, is important in this regard.

The capacity of the institutional apparatus of the IPCC to both separate and integrate science and politics in the assessment process may be seen as a function of the balance characterising the current rules of procedure between governmental control at the upper decision-making levels and scientific control at the lower decision-making levels of the system. While this balance to some extent has evolved through experience and learning during the course of the process, the WG and panel chairs seem to have been instrumental in the incremental development of formalised rules and procedures whereby this balance is achieved. That is, the balance between governmental and scientific control characterising the formal rules of procedure of the IPCC do not seem to be the result of a coincidental evolution. Rather, this balance seems to be the result of a deliberate design carefully negotiated by the WG and panel chairs.

The Symbolic Value of Rules of Procedure

During the course of the IPCC process, the rules of procedure have acquired a strong symbolic value, mainly as the prime indicator of the scientific integrity of the IPCC. In this capacity, the rules of procedure represent a tool for several purposes. First, they have become a major tool for some delegations to question the scientific legitimacy of the process, and hence the reliability and validity of the findings and conclusions brought forward in the assessments. In this function, the rules of procedure, or rather the extent to which the rules of procedure actually have been followed, become the main object of scrutiny, not the findings and conclusions themselves. Thus, the scientific validity of findings and conclusions are only indirectly questioned. The argument, then, may go along the lines that, "these findings may well be true, but since they have not come about in the correct manner – and we cannot, therefore, be *certain* that they are true – they should be deleted". In some cases, the question of scientific validity is not even addressed; it is simply argued that whatever is subject to discussion represents a violation of the IPCC's rules of procedure and should therefore be deleted. Second, mirroring this rhetorical employment of the rules of procedure as a lever against the scientific validity of conclusions and findings, the rules of procedure have also become a tool particularly for the WG and panel leadership to defend the same conclusions and findings of the reports. The argument then would go along the lines that, "on the contrary, we can demonstrate that these findings have come about in an absolutely correct manner, and do not represent a violation of IPCC rules and procedures and should therefore, unless they are demonstrably false, be included in the report". During the course of the IPCC process, such curious "meta-discussions", which actually may become quite time consuming and complicated, seem to have become the rule rather than the exception in WG and panel plenary sessions.

The discussion which took place at the WGI plenary session in Maastricht in 1994 over one particular paragraph in the SPM for the 1994 report exemplifies the symbolic and rhetorical dimension of the rules of procedure. The discussion concerned paragraph 2.3.2 of the SPM (numbered 3.3.2 in the draft SPM), which investigates emission profiles that would lead to a stabilisation of the atmospheric concentration of CO_2. The question concerned whether this paragraph was an assessment of published and reviewed literature, or whether it actually was based on studies commissioned by the IPCC that were not yet published and properly peer-reviewed, in which case, the paragraph, according to the view of some delegations and NGOs, should be deleted. The proposal to delete the paragraph in question was put forward by the Global Climate Coalition (GCC), a US-based fossil fuel industrial lobby group, and was supported by the major oil-producing countries, the Kuwaiti delegation in particular. The following statement from the GCC sparked the discussion:

> It appears that the IPCC actually commissioned the work reported in this section. It is also known that the basic paper from which the results are obtained has not yet completed the peer review process. In our view, the IPCC was formed to *assess* the literature, not to *create* the literature. It is also our view that only peer reviewed papers should be included in the IPCC process. In order to protect the integrity of the IPCC process, we

strongly suggest deleting this section. (Statement made by the GCC at the fourth WGI plenary session in Maastricht, 13–15 September 1994, emphasis in original.)

The WG and panel leadership emphasised that while these studies were new and preliminary, they had indeed been subject to a proper peer review and were, although they had not yet been published, available to the reviewers of the IPCC report. Moreover, they emphasised that the workplans of WGI were approved by the panel plenary in Harare in 1992, and re-approved by the panel plenary in Geneva in 1993 and, hence, that the inclusion of the paragraph did not represent a violation of the rules of procedure. It was furthermore emphasised that, in order to be properly updated, this was the procedure that had to be followed by the IPCC. One delegation also confirmed that the review of the report itself was completed, as well as the peer review of the referred literature upon which the section in question was based.

The Kuwaiti delegate did not find these remarks convincing. In particular, he objected to the detailed treatment of CO_2 in contrast to a less detailed investigation of other greenhouse gases, and maintained that the paragraph could lead policymakers to erroneous conclusions. Moreover, he maintained that the terms of reference were not followed and that the inclusion of this paragraph would be a violation of the plan of action approved by the panel plenary. On this basis, the Kuwaiti delegate found it "extremely necessary" to delete the section. By the leadership it was, however, pointed out that CO_2 required particular attention due to the complex nature of the lifetime of atmospheric CO_2, and that similar, although not equally detailed, investigations were made for methane (CH_4) and nitrous oxide (N_2O) in subsequent sections of the SPM (personal observation).

The position of the WG and panel chairs to retain the section received an overwhelming support from the majority of delegations. After some discussion, the panel chair stated that "in this discussion, I do not sense a *scientific* controversy", and ruled that the dissenting view of this minority of delegations could, if they so wished, be recorded in a footnote, but that the paragraph would be included in the report since this was the expressed view of a majority of the delegations at the meeting (cited from personal notes from the meeting). Ultimately, a compromise was agreed upon, which implied that the WGI chair would draft a "preface" to the section, in which the particularly detailed investigation of CO_2 was properly explained (see paragraph 2.2.3 of the SPM to the 1994 Special Report).

The rhetorical potential associated with rules and procedures implies a significantly reduced flexibility in the provision of reports. Rules and procedures cannot be violated during the assessment process. This aspect is also demonstrated by the procedural discussions which took place at the same WGI plenary meeting (Maastricht).

Procedural questions were subject to particularly heated debates at this plenary session of WGI because the final draft of the bulk report had not been distributed within the time limits required by the rules of procedure. The rules of procedure state that "[a] final draft taking into account country comments should be distributed to all IPCC participating countries and organizations not less than three weeks prior to a Working Group Plenary meeting called to accept the report..." (item 5 of the 1993 rules of procedure). At the Maastricht meeting, some delegations had not received the final (third) draft at all, prior to the meeting.[5] This issue was brought up by Kuwait, and was subject to both formal and informal discussions during the first day of the WGI plenary meeting. The Kuwaiti delegation also made attempts to utilise this clear violation of rules of procedure as a lever for increased influence on the drafting of the SPM. This is particularly indicated by their statement to the WGI plenary:

- Having a discussion on the draft SPM without the availability of the third version of the underlying report is a clear violation of the rules of procedure.
- Kuwait shall have significant comments to make on the SPM.
- *Depending on the outcome of how our comments are treated*, my Government then will decide what position it will take regarding this unfortunate situation. Having said that, my delegation reserves the right to object at the IPCC plenary in Nairobi to the effort at this WGI Plenary session either to approve the policy maker summary (SPM) or to accept any underlying report. (Statement by Kuwait at the fourth WGI plenary meeting in Maastricht, 13–15 September 1994, emphasis added.)

In the meeting, the Kuwaiti delegation stated that if a compromise was achieved, they would accept the procedures and not reopen the discussion at the subsequent plenary session to be held in Nairobi in November in 1994. If not, however, they would reopen the issue in Nairobi. The issue was resolved by a compromise achieved in informal discussions between the leadership of WGI and the Kuwaiti delegation during a lunch break. At the plenary session in Nairobi, the SPM as well as the underlying report was accepted without further discussion.

[5] It should perhaps be mentioned that all delegations did have copies of the second draft of the underlying report and that changes made in the reports at this stage of the assessment process usually are largely editorial in kind. The failure to distribute the final version of the report was, however, a clear violation of the rules of procedure and was acknowledged as such by the WG and panel chairs.

The situation that arose at the Maastricht meeting was subsequently pursued, however, by both the US State Department and the GCC (*Nature*, vol. 371, October, 1994). Moreover, the situation that arose at this meeting put the lead authors in a difficult position. One of the contributing scientists is reported to have said that "it really upset the lead authors, who felt that policymakers were making changes to the conclusions that were not really based on the supporting material" (*Nature*, vol. 371, 6 October 1994).

These debates indicate the complexity of procedural questions, particularly because of the symbolic value the rules of procedure have attained in the IPCC process, and the underlying political concerns and strategies. Given this rhetorical dimension of rules and procedures, their *formalisation* is important. Interventions of this kind are likely to be pursued regardless of whether the rules and procedures are formalised or not. While a formalised set of rules and procedures to some extent gives strength to this line of argument, it also, by the same logic, provides the WG and panel leadership with an instrument whereby to handle such interventions.

The Formalisation of Rules of Procedure

The adoption of rules of procedure is a matter for the panel plenary to decide. Any procedural change must be accepted by consensus by governments in the panel plenary. In this regard, governments have a substantial amount of control over the IPCC's mode of operation. Some governments are sceptical towards procedural changes in general, and particularly procedural changes that may serve to restrict governmental control over the assessment process at the upper decision-making levels of the IPCC system. Curiously, any proposal to this effect is usually met by a concern for the scientific integrity of the IPCC. Any proposal for procedural change must, therefore, be carefully prepared and negotiated in the panel plenary.

The political nature of the forum where the rules of procedure for the IPCC are adopted could imply that the current rules of procedure reflect a negotiated aggregation of governmental interests. The nature of the rules of procedure themselves, however, and especially the balance achieved between governmental and scientific control at different decision-making levels, does not support this interpretation. Several procedural decisions in the panel plenary do not reflect "governmental" interests, at least not if governmental interests are defined as having as much governmental control over the assessment process as possible. Rather, the current rules of procedure seem to constitute the result of a conscious design, carefully prepared and negotiated by the WG and panel leadership. Two examples may serve to illustrate this point.

At the ninth plenary session in 1993, the IPCC rules of procedure underwent a major revision. One of the items on the agenda concerned the selection of lead authors. The sixth session of the IPCC bureau recommended that, first, teams of lead authors should be selected in which at least one came from a developing country. Second, the bureau recommended that the teams of lead authors should be composed so that differences of scientific opinion would also be represented, thereby providing a built-in mechanism to deal with controversies that might arise (report of the ninth plenary session). During the plenary discussion, several questions related to this matter were raised, of which the most important concerned the extent to which the plenary should be involved in the selection of lead authors. Some governments maintained that especially given the increased emphasis on the representativity of the teams of lead authors in terms of both geographic balance and differences of scientific opinion, the choice of lead authors made by the bureau of each WG should be subject to plenary approval. Had this proposal been adopted, governments would have acquired a substantial influence over the selection and composition of the teams of lead authors, and the operational autonomy of the scientific core of the WGs would have been reduced accordingly. Moreover, this would have implied an extremely cumbersome and time-consuming procedure, which they might not even be able to follow in practice. Given the instrumental dimension of rules of procedure as a tool for pursuing political strategies, it is important not to adopt rules of procedure that may prove to be impossible to implement. The panel leadership bears a major responsibility for ensuring that "impossible" rules of procedure are not adopted, since some delegations may regard the adoption of "impossible" rules of procedure as an instrument in a delaying strategy. This proposal was, therefore, insistently opposed by the WG and panel chairs. Finally, after having discussed this issue several times during the course of the meeting, these delegations agreed to the procedure whereby lead authors are selected by the bureau of the WGs on the basis of lists of nominations made by governments but where governments do not have any direct control over the actual selection (personal observation). It is interesting to note that this procedure was already in operation, and had been since 1988. Still, its formal institutionalisation required a fair amount of negotiation and persuasion in the panel plenary.

At the eleventh plenary session in 1995, the bureau of the panel put forward a proposal for consideration by the plenary to the effect that the IPCC should provide a third category of reports – Technical Papers – in addition to Assessments and Special Reports. The proposal was motivated by a concern for the IPCC's ability to respond to short-term requests for scientific and technical information and advice in a timely manner. While the bureau acknowledged the process whereby IPCC reports are provided as the basic strength of the IPCC process, its time-consuming nature was also recognised as a barrier to effective communication with the COP (report of the eleventh plenary session; Discussion Paper prepared by the Chairman of the IPCC, IPCC-XI/Doc. 7). The provision of Technical Papers, therefore, was proposed to follow a simpler procedure, notably not being subjected to a line-by-line approval by either the WG or the panel plenaries. In particular, the provision of Technical Papers would follow a procedure where the reports should: (i) be based on material already in the IPCC assessments and special reports; (ii) be prepared by a team of lead authors selected according to the rules of procedure; (iii) be submitted to a concurrent expert and government review with an additional government review after revisions from the first round; (iv) be finalised by lead authors in consultation with the IPCC bureau in the role of an editorial board; and (v) if deemed necessary by the IPCC bureau, include an annex reflecting differing views if these were not considered to be adequately reflected in the paper (report of the eleventh plenary session, agenda item 5.9).

In this proposal, the necessary balance between governmental and scientific control is taken into account in terms of the role ascribed to the panel bureau. Some governments, however, expressed serious concern that the plenary was not assigned a role in this process. These delegations maintained, first, that the topics for which the IPCC should provide technical papers should be decided by the plenary, and second, that the technical paper itself should be subject to approval, or at least acceptance, by the plenary. Had these proposals been adopted, however, technical papers would, in effect, be no different from special reports, and the IPCC's ability to produce scientific information and advice in a timely manner would not be increased. The issue was raised, postponed and re-raised during the course of the meeting, and finally agreed upon by the plenary as one of the last unresolved items on the agenda (personal observation). It was, however, specifically decided that the initiation of Technical Papers was to be discussed and agreed upon by either the bureau or the panel plenary. Thus, no other IPCC body, such as, for instance, a WG plenary, can decide that the IPCC shall provide a technical paper on a specific topic.

When the panel leadership initiate procedural changes, the issue has first been discussed in the IPCC bureau, and usually also in the bureau of each WG. The IPCC bureau is composed of members representing a geographic balance, and can be assumed to be representative of the major (conflicting) interests of IPCC member states (see chapter 6, section 6.4.3). The discussion in the bureau gives the panel leadership an impression of which arguments the proposal may meet in the plenary. Before the issue is raised at a plenary meeting, it is also discussed further internally within the WG and panel leadership, who also, individually, approach governments informally.

The issue is then pursued in a plenary session by the WG and panel leadership (personal communication). In the plenary, issues of procedural change are carefully and incrementally negotiated. That is, the issue is discussed, postponed and re-raised throughout the meeting, permitting the leadership to have broad, informal consultations with delegations, in "corridor" conversations, or at informal side meetings, where the concerns of reluctant and sceptical delegations can be considered and, where possible, accommodated. In this manner, the WG and panel leadership usually get it the way they want (personal communication).

The WG and panel chairs serve an important role in the negotiation of procedural changes – that is, in ensuring and facilitating the process whereby procedural arrangements and changes are adopted by the panel plenary. Their most important function, however, is linked to institutional design. While some delegations are sceptical towards procedural changes, most delegations welcome proposals that may serve to facilitate and enhance the effectiveness of the process. In the adoption and enforcement of procedural arrangements that may serve to enhance the effectiveness of the process, therefore, the WG and panel chairs are usually supported by the majority of delegations. Delegations themselves usually do not, however, initiate procedural changes to this effect. Thus, in the development of the institutional design of the IPCC process – the development of a system of institutional arrangements designed to enhance the effectiveness of the process – the WG and panel chairs bear a main responsibility. In this regard, procedural management not only constitutes a necessary capability to perform leadership roles successfully, it is a leadership function in itself, one to which the WG and panel chairs seem to have provided a substantial contribution.

Moreover, while the current rules of procedure have been developed incrementally during the course of the process, procedures for the provision of assessments seem to have been informally established – by precedent – at a relatively early stage of the process (largely during the first assessment), particularly as a function of WGI's mode of operation. For instance, one of the key features of the IPCC process – namely the development of Summaries for Policymakers and Executive Summaries, and the procedures whereby the summaries are approved by the panel in WG plenaries – was actually established by the WGI leadership during their work with the First Assessment Report (personal communication). This procedure implies that governments only approve the summaries, not the actual assessment reports themselves. Moreover, the lead authors of the WG play an important role in the WG plenary's approval of the summaries. The participation by lead authors in WG plenaries is the key component of this decision-making level's important function as an arena for an interactive dialogue between scientists and policymakers, which has significant implications for the nature and functioning of the IPCC process. In this regard also, therefore, the leadership of WGI has had a major impact on the design and functioning of the IPCC process.

Given this informal establishment of procedures, procedural changes at later stages of the process have to a very large extent been formalisations of procedures already in operation, rather than actual changes in the institution's mode of operation. Still, however, given the potential to utilise the rules of procedure as an instrument in political strategies, their formalisation has required a substantial amount of diplomacy.

The incremental nature of the process whereby the rules of procedure of the IPCC have been developed and formalised implies, however, that no radical changes in rules of procedure have been either proposed or made overnight. The rules of procedure have been developed step-by-step in an incremental process based on experience and learning. In this gradual formalisation of rules of procedure, however, the IPCC leadership seems to have played a significant role in achieving the balance between governmental and scientific control currently characterising the IPCC process. In this effort, they seem to have contributed substantially to the current design of the IPCC process and to the capacity of the institutional apparatus of the IPCC to combine a separation of science and politics with the provision of arenas for an interactive science–policy dialogue.

7.2.5 Boundary-role Leadership in the Provision of Communicative Links

The capacity of the institutional apparatus of the IPCC to both separate and integrate science and politics is a function of an institutional design that serves to separate the scientific arenas from the arenas where scientists and policymakers interact. As pointed out in chapter 6, the three decision-making levels are institutionally separated and there are relatively few institutional arrangements that serve as links. Communicative links between the decision-making levels of the institution are, however, important for the functioning of the system as a whole, and seem mainly to be provided by individual agents. In this regard, the capacity of the institutional apparatus of the IPCC to both separate and integrate science and politics is also a function of the capacity of individual agents within the system to provide this communicative function.

While lead authors play a major role both within the scientific core (in the development of assessments) and in the WG plenary (in the negotiation of the SPMs), it is only the WG and panel chairs that participate at all decision-making levels, from scientific core to panel plenaries.[6] These actors, therefore, bear the main responsibility for the provision of communicative links within the IPCC system. Moreover, the panel chair is mainly responsible for the provision of communicative links between the IPCC and the political community as represented by the INC/COP.

The panel chair, Prof. Bert Bolin, states that he sees it as one of his main tasks and main challenges as chair of the IPCC to communicate effectively the needs of policymakers to the scientific community and the nature and limits of research input to the political community (personal communication). With regard to the former, he has devoted a lot of energy to stimulating the scientific community to employ a comprehensible language in the SPMs (not "too technical") and to include, to the greatest extent possible, all viewpoints in the assessments made. Now and in the future, he maintains, the strength of the IPCC will depend upon its ability to include all (opposing) scientifically substantiated viewpoints in the assessments (personal communication).

In his introductory address to the second drafting session of WGI for the 1995 Assessment Report, Bolin emphasised the importance of structuring the presentation of research findings so that they may serve as an actual aid to policymakers in their task. He indicated that in this regard, the SPM of the 1994 Special Report was perhaps "a little bit too technical", and urged the lead authors to "take one more step into the context where the knowledge is to be used" (cited from personal notes from the meeting). Moreover, the presentations by the teams of lead authors on their progress with their respective chapters of the assessment was followed by Bolin's question, "what is your message to policymakers?" The answer he got was usually followed by the comment, "if that is your message to policymakers, you should say so!" (cited from personal notes from the meeting).

In the same manner that the WG and panel chairs, in a sense, "speak the case of policymakers" at meetings within the scientific core, they represent the scientific community in panel plenaries, where scientists usually do not participate. Their role as representatives of the scientific community seems to have been particularly important in relation to the panel plenary's approval of the Synthesis Report, although it is difficult to pinpoint exactly what the role implied. For instance, in the panel plenary's discussion and approval of the 1995 Synthesis Report, the function of the WG and panel chairs was closely linked to their general overview over IPCC reports – that is, their detailed knowledge about the substantive content of the SPMs of all the three WGs as well as detailed knowledge about the substantive content

[6] As a matter of fact, it is only the panel chair himself who participates at all decision-making levels of *all* the WGs. Each WG chair only participates at all the decision-making levels of his own WG.

of the draft synthesis report itself. The synthesis report was supposed to be based on the conclusions of the three underlying reports, but given its explicit aim to aid policymakers in their discussions and assessments of what a "dangerous" anthropogenic interference with the climate system would be, it drew selectively on the material of the underlying reports. Thus, at every little juncture in the discussion over specific formulations in the synthesis report, the WG and panel chairs had to know what the SPM of the relevant WG said on the subject, and where it could be found. In a sense, therefore, their guidance and leadership role in the panel plenary's discussion is only to a little extent reflected in what they said and did at the meeting. Rather, their leadership role lies in their intimate knowledge of the scientific basis of the report, which is a condition for their ability to maintain their "grip" on the discussion and provide general guidance. Thus, their comments and proposals in the discussion of each paragraph (and sentence) of the synthesis report may actually seem insignificant when viewed separately. Taken together, however, their contribution to the discussion seems to sum up to a rather significant guidance of the debate and seems to have served as the major mechanism whereby the correspondence between the synthesis report and the underlying reports was ensured (personal observation).

The synthesis report is organised in sections corresponding to the fields of the three WGs. Thus, in the discussion of the various sections of the synthesis report, the panel chair is assisted by the chair of the relevant WG in his chairmanship of the discussion. While WG and panel chairs clearly do not have a veto power similar to that of lead authors in WGI plenary sessions, they can co-ordinate their chairmanship and assist each other in ensuring the correspondence between the reports. The scientific authority of the WG and panel chairs seems generally to be respected by delegates, and their rejection of proposals for change that they view as inconsistent with the underlying reports also seems to be respected on this basis. In cases where agreement proves difficult, they use language from the SPMs, insert clarifications and definitions in footnotes to the text, or, as a last resort, record dissenting views.

Their detailed knowledge of the scientific basis of the synthesis report, therefore, seems to have enabled the WG and panel chairs to establish firm limits to the extent of possible modifications to the text, and thereby to some degree provide the function that the veto power of lead authors serves in WG plenaries. In this manner, the role of the WG and panel chairs in panel plenary sessions seems to have served to facilitate the development of consensus, as well as to ensure consistency between the various reports.

By communicating the needs of policymakers to scientists in their work with the development of assessments within the scientific core and by acting as representatives of the scientific community at decision-making levels dominated by government officials, the WG and panel chairs thus seem to have contributed substantially to the provision of the communicative links between decision-making levels that are necessary for the functioning of a system in which these parts of the process are institutionally separated.

The IPCC's communication with the political community, as represented by the INC/COP, is institutionalised in the "Joint Working Group" (before 1995, known as the "IPCC–INC Joint Working Party") (see also Agrawala, 1998b). Established on Bolin's initiative in 1993, this is where the senior officials of the IPCC and COP meet. In addition, Bolin reports to the INC/COP on a regular basis. In these meetings, two types of information seem to have been emphasised: general, and to some extent practical, information about the IPCC (when they will discuss various issues, when their reports are due, which topics are addressed in forthcoming reports, etc.); and provisional summaries of the main conclusions of forthcoming reports. In addition, the chairman has on several occasions addressed questions related to the appropriate boundaries between science and politics in terms of the delineation between IPCC and INC/COP activities, as well as suggesting possible policy implications and providing cautious policy advice. Bolin's communication with the INC/COP, therefore, does not seem to have served merely to inform policymakers about IPCC activities. These meetings have also served as an opportunity to suggest and discuss the delineation between IPCC and INC/COP activities and to suggest possible policy implications and provide cautious policy advice.[7]

After having participated at INC/COP sessions, the panel chair reports back to the IPCC bureau. After his participation at the second session of the INC in 1991, for instance, Bolin communicated to the IPCC bureau views expressed by governments at the INC meeting concerning the question of whether the IPCC should assess socio-economic implications of climate change, which was a hot and debated question within the IPCC at that time (letter from Bert Bolin to members of the IPCC bureau, dated 8 July 1991).

In these respects, the panel chair does seem to have served in a boundary role where he attempted to "match" the supply of and demand for research input to the policymaking process. What this boundary role has implied beyond this "reporting" function is, however, hard to establish. As discussed above, the functions are in themselves subtle and hard to pinpoint. Moreover, some of the meetings in which communicative links between the

[7] In his address to the ninth plenary session of the INC in 1994, for instance, Bolin cautiously advised policymakers to start preparations for more ambitious goals of emissions reductions to stabilise atmospheric GHG concentrations at or below pre-industrial levels. He also brought up and expressed his view regarding the choice of time horizon for GWPs although this, as he stated, "is not a scientific matter" (Report to the ninth plenary session of the INC by Bert Bolin, 7 February 1994).

IPCC and the political community are established are informal in nature and thus without official records (for instance, the meetings of the Joint Working Group).

The panel chair has undoubtedly been assisted by some very competent people in the provision of boundary-role leadership both within the framework of the IPCC itself as well as in the IPCC's communication with the political community both at the national and international level. In particular, Dr. Robert Watson's name is inescapable in this regard. Within the IPCC system, Watson has operated at all decision-making levels: He was a contributing scientist in WGI to the First Assessment (1990) and the Supplementary Report (1992); he served as chairman of WGII from 1992 to 1997; he has served as US delegate to all WG and panel plenaries (other than WGII sessions after 1992); and he is the newly elected chair of the panel (1997). Moreover, with his Ph.D. from London University in 1973 on the gas-phase kinetics of chlorine, bromine and fluorine, and his experience as programme director at NASA, leader in the international WMO/UNEP sponsored ozone assessment process, associate director for the environment at the White House's Office of Science and Technology, director of the environmental department at the World Bank, and chairman of the Scientific Assessment under the Convention on Biodiversity, Watson has operated in the interface of science and politics his entire career. He has been characterised as "a hard-driving, consensus-building science manager", a "bureaucratic entrepreneur" and a "network builder" (*Science*, vol.277, September, 1997). In his many capacities, he has a very broad network encompassing both the scientific and the political communities, and he is reported to have a very "energetic approach" to science (*Science*, vol. 277, September, 1997).

With his broad professional background and his personal energy and skill, Watson is most likely to have served a boundary-role leadership function in the IPCC process, although the data in this study are too limited to document and specify this role further. However, in his capacity as newly elected chair of the IPCC at the panel's thirteenth plenary session in 1997, for instance, he addressed the question of very significant procedural changes in the panel's preparation for the third assessment in an impressively effective manner. The issue was very well prepared: A paper drafted by Watson in which the necessary procedural changes were discussed had been circulated for review three times before the meeting, and in the final version presented to the thirteenth plenary session, all decision points were made explicit. Perhaps most importantly, the paper was drafted with supplementing comments to all decision points where delegations could have their views recorded, but which, in themselves, were not subject to (line-by-line) approval. Thus, all points of view could be channelled towards and recorded in the comments without time-consuming and cumbersome word-by-word negotiations. In this manner, the panel plenary managed to make some fifty decisions on procedural issues in one and a half hours, which is quite extraordinary (personal communication with Bert Bolin; Decision Paper from the thirteenth panel plenary session). Moreover, reportedly the IPCC "will try to refine the scientific consensus further, pushing assessments toward the edge of – but not into – policymaking", under the chairmanship of Dr. Watson (*Science*, vol. 277, September, 1997). His multiple roles, broad professional background and access to a wide network encompassing both the scientific and the political communities are his strengths in this effort.

This discussion indicates that WG and panel chairs have served important functions in boundary roles within the IPCC system itself, as well as in the IPCC's communication with the political community. These individuals have clearly played an influential role in the incremental development of the IPCC's institutional design. Their functions in the provision of both internal and external communicative links also seem to have been significant, although this role is more subtle and difficult to pinpoint in its precise nature. The provision of communicative links between the IPCC and the political community is also likely to have taken place at arenas external to the IPCC framework – at the national level, at workshops and conferences, etc. – and hence beyond the scope of this analysis.

7.3 Institutional Conditions for Leadership Performance

To what extent are these modes of leadership linked to specific institutional arrangements? Leadership performance may be seen as a function of institutional arrangements in at least two regards: First, in the sense that the performance of leadership functions depend on the extent to which institutional arrangements provide the necessary mandate for this type of behaviour. Also, institutional arrangements may to a varying extent provide individual actors with the necessary incentives and flexibility to perform in this role. Second, in the sense that institutional arrangements, rules and procedures to a varying extent specify criteria for recruitment into leadership positions so that the recruitment of individuals with capabilities to serve a leadership role is ensured. However, while the provision of leadership functions is determined by institutional arrangements in these respects, the extent to which and the manner in which individual actors seize and realise this potential may vary. Thus, the provision of leadership is (partly) a function of institutional arrangements, but since other factors also affect the extent to which leadership is actually provided (for instance the personal capabilities and the interpersonal skills of the candidate), the provision of leadership constitutes an intermediate variable which may have an independent impact on the outcome of the process.

7.3.1 A Mandate for Leadership?

Institutions of the magnitude of the IPCC will always provide a mandate for leadership in key positions such as WG and panel chairs and vice chairs, secretariat functions etc. Beyond this absolutely necessary mandate for leadership, the institutional design of the IPCC provided a mandate for leadership in two important respects: First, in the role assigned to lead authors in WG plenary sessions; and second, in the mandate for boundary-role leadership implicit in a system where a few key positions constitute the main link between institutionally separated decision-making levels.

A basic condition for the leadership function provided by lead authors in the transformation of the knowledge base is the central role ascribed to lead authors in the WG plenary's approval of the summaries for policymakers and executive summaries. The participation of lead authors in plenary sessions serves to provide an arena where lead authors may perform this role, and is a key element in the provision of an arena where scientists and government officials can engage in an interactive dialogue. Moreover, this mandate for leadership is supplemented by institutional devices that are instrumental for facilitating and providing incentives for the development of consensus in the WG's negotiation of the summaries. This is particularly true with regard to the procedures which provide lead authors with a veto against substantive changes to the summaries that could distort the correspondence between the summaries and the bulk report upon which they are based. This arrangement may have served as an incentive for governments to modify their proposals in order to increase their potential influence on the text and may in this respect be seen as an institutional device that facilitates the development of consensus. In a sense, this arrangement defines the "limit" to governmental influence on the formulations of the summaries. Moreover, it also seems as if governments are not very keen on having their dissenting views recorded in a footnote. Even in the situation that arose at WGI's fifth plenary session in Madrid, where some governments insisted that the word "preliminary" be added to the text of the SPM and they were given the opportunity to record this view in a footnote, they ultimately withdrew their dissent (see chapter 6, section 6.3.2, footnotes 6 and 7).

Thus, the role assigned to lead authors in the approval of the summaries defines a framework within which consensus can be developed as well as providing incentives for both scientists and governments to actively contribute to this development. This institutional device, therefore, does not reward rigid and inflexible positions with increased influence. Rather, parties are dissuaded from adopting a negotiation strategy directed towards driving negotiations into deadlock situations in the hope that their inflexibility will increase their influence. In this regard, the role of lead authors in WG plenaries is parallel to the role of the committee chairmen in the drafting of the "Single Negotiating Text" (SNT) in the Law of the Sea negotiations (see, for instance, Buzan, 1981; Skodvin Hegdal, 1992). In the Law of the Sea negotiations, the introduction of the SNT procedure marked a shift from passive to active consensus in the sense that the method provided governments with incentives to engage actively in problem solving and consensus development. Before this procedure was introduced, there were no incentives for governments to take the first step out of deadlock situations (Buzan, 1981).

Thus, the institutional arrangements of the IPCC[8] constitute an important condition for the potential of lead authors to serve leadership roles in the transformation of the knowledge base. However, the extent to which lead authors have converted this potential into actual leadership performance also seems to depend upon their personal ability to develop text that integrates as much as possible the concerns of both scientists and government officials. Thus, while these institutional arrangements constitute a necessary condition for the leadership functions provided by lead authors in WGI plenaries, they are not sufficient to *ensure* that leadership actually is provided.

The question of whether the boundary-role leadership served by WG and panel chairs is conditioned by an institutional mandate for leadership is more complex. First, WG and panel chairs seem to have served a leadership function directed towards the design of an institutional apparatus capable of both separating and integrating science and politics. In this regard, the question of institutional conditions for the provision of leadership concerns the extent to which leadership behaviour directed towards the development of institutional arrangements *in itself* is conditioned by already existing (or non-existing) institutional arrangements. Second, the question concerns the extent to which the institutional design thus developed provides a mandate for (further) leadership behaviour.

The leadership function served by the WG and panel chairs in the development of explicit rules of procedure for the IPCC process is in a sense conditioned by the lack of formal rules of procedure when the IPCC process started in 1988. While this lack did cause problems on some occasions (for

[8] The role of lead authors in WG plenaries is, as a matter of fact, not formally established in the rules of procedure of the panel. This mode of operation was adopted by WGI during the first assessment process, and has gradually been transferred to all WGs (particularly after the 1992 restructuring of the panel). In this regard, the system is based upon *informal* positions of authority.

instance in the old WGIII as well as in the adoption of the first assessment), it also provided the WG and panel leadership with an opportunity to develop an institutional design that carefully accommodated the needs of the IPCC process, particularly with regard to combining a high level of scientific autonomy and integrity with a high level of science–policy integration. Thus, the lack of formal rules and procedures at the start of the process permitted an incremental development of an institutional design based on experience and learning, to whose development the leadership served by WG and panel chairs seems to have contributed substantially.

While the lack of formal rules and procedures when the IPCC started its work certainly constitutes an important condition for the role of the WG and panel leadership in the development of the IPCC's institutional design, it hardly qualifies as a "mandate" for leadership. Rather, it provided the WG and panel chairs with an opportunity to serve a leadership role directed towards this function which they seem to a large extent to have seized and utilised to develop an institutional apparatus designed according to the specific needs of this process in order to enhance its effectiveness. This potential might not have been realised in this manner, however. On the contrary, the lack of formal rules and procedures during the initial phase of the process served to increase the vulnerability of the process to undue politicisation. Moreover, the lack of formal rules and procedures could have been used as an opportunity to enhance the political control over the panel's work. Thus, while the institutional arrangements – or lack thereof – of the IPCC process during its initial phase provided the WG and panel chairs with the necessary "room" to serve a leadership role directed towards institutional design, the manner in which this feature was utilised as an instrument to enhance the effectiveness of the process is largely explained by the efforts and capabilities of the WG and panel chairs.

Implicit in the institutional design of the IPCC thus developed, there is considerable scope for leadership behaviour directed towards the provision of communicative links between the institutionally separated decision-making levels. In this regard, WG and panel chairs contributed to the development of an institutional design which not only serves to enhance the capacity of the institutional apparatus to both separate and integrate science and politics, it also provides a mandate for further leadership behaviour. Apart from this mandate, however, this mode of leadership is to a very little extent conditioned by institutional arrangements. The specific tasks implied in this role have not been made explicit and the leadership of the IPCC have no specific institutional devices at their disposal that may serve as instruments to facilitate their provision of the communicative function. Moreover, beyond a general objective of providing links and communicating various forms of information between the decision-making levels of the institution, the leadership function is not associated with any specific objective (parallel, for instance, to the objective associated with the role of lead authors in WG plenaries of developing consensus on report summaries). The substantive content of this mode of leadership in terms of explicit behaviour, therefore, is heavily dependent upon what the individuals operating within the role make of it. In this regard, this mode of leadership is formed by the individuals occupying the role.

While this mode of leadership to a very little extent is conditioned by institutional arrangements, it still serves to illustrate the interplay characterising the relationship between institutional arrangements and leadership behaviour. The differentiation of institutional arrangements according to function characterising the IPCC implies a system with no or few institutional linkages between decision-making levels. This system produces a particular demand for leadership directed towards the provision of communicative links between decision-making levels and the functioning of the system depends upon the extent to which this mode of leadership is provided by agents operating within the system. It is important to recognise, moreover, that the design of a system where decision-making levels are separated is strongly linked to the dilemmas involved in the design of a system capable of ensuring both scientific integrity and science–policy interaction and involvement. It is precisely the institutional separation of decision-making levels with different functions in the assessment process that contributes to the capacity of the IPCC system to combine scientific integrity with science–policy integration. Thus, the institutional separation of decision-making levels may be regarded as a means whereby the capacity of the system to both separate and integrate science and politics is enhanced. We may assume that a system not characterised by this functional differentiation would be less capable of combining scientific autonomy and science–policy integration and hence less effective. However, if the functional differentiation characterising the IPCC system were not supplemented by the provision of communicative links by agents operating within the system, the effectiveness of the science–policy dialogue would also be reduced. In this regard, the impact of institutional arrangements strongly depends on the provision of boundary-role leadership just as the scope (demand) for and provision of boundary-role leadership strongly depends on institutional arrangements.

7.3.2 Leadership Recruitment

The provision of leadership may be regarded as a function of institutional arrangements also in the sense that recruitment criteria determine the extent to which actors with the necessary capabilities to perform leadership roles are recruited to key positions.

The main capability required to perform a leadership role on the basis of a position as lead author is substantive knowledge in the form of scientific qualifications and experience. As we have seen, formal scientific qualifications constitute the main criteria for the selection of lead authors to WGI. Thus, all lead authors in WGI seem to hold a potential to perform a leadership role in the provision of assessments. Our discussion also shows, however, that scientific brilliance is not enough to be a good lead author in WGI. The extent to which a lead author succeeds in converting his leadership potential into actual leadership also seems to depend upon two other factors: (i) his or her ability to contribute to the development of procedures that serve to generate contributing authors' trust that their comments are taken into account and incorporated into the assessments on the basis of their scientific merit; and (ii) his or her ability to induce co-operation among participants and to handle the disputes and conflicts that may arise in the course of the process. Thus, the interpersonal and human skills of the lead authors are also important for their capability to carry out a successful leadership. The significance of the personal dimension associated with the role is illustrated in the difficult situation which arose after the lead author shift in WGI in 1990.

It is difficult to implement a recruitment system where these personal capabilities are emphasised. Thus, the criteria guiding the recruitment of lead authors cannot be specified to the extent that performances of leadership are ensured through institutional arrangements alone. This suggests, therefore, that the extent to which leadership has been provided in the assessment process of WGI is not merely a function of institutional arrangements. While institutional arrangements constitute a necessary condition for lead authors' potential to serve a leadership role, they are not sufficient to ensure that leadership is actually provided.

The leadership functions provided by the WG and panel chairs in the transformation of the knowledge base seem to an even lesser extent to be ensured through institutional arrangements. Given that the IPCC is a UN body, the selection of individuals to these positions is subject to negotiations among governments. The panel chairman is elected by the panel plenary on the basis of nominations made by member governments to UNEP and WMO. Although formally elected by the panel plenary, WG chairs are determined by the distribution of key positions among governments. The extent to which individuals with the (personal and scientific) capabilities to perform a leadership function are recruited to these positions is, therefore, largely determined by governments at the national level.[9] This is very well illustrated in the selection of WG chairmen to WGs II and III before 1992, where it appears that the political and administrative positions of the selected chairmen at the national level were given more emphasis than their scientific merit and qualifications. This seems to be particularly true with regard to the selection of the chairman of WGIII, who also served as head delegate of the US delegation to INC sessions. The role of the WGII chairman in the conflict between WGs I and II over the paleo analogue method also suggests, however, the political function associated to his chairmanship (see chapter 6, section 6.4.4). The selection of more or less "political" actors into these positions by national governments before 1992, may be a reflection of the political significance ascribed to WGs II and III during this phase of the IPCC process. Moreover, it seems likely that some of the increased politicisation these WGs experienced during this phase as compared to WGI may be explained by these nominations.

Thus, while governments are encouraged to "ensure the scientific-technical integrity and credibility of the IPCC by nominating relevant experts for participation in the work of the Panel" (report of the eighth plenary session, agenda item 6.16.1), there are no formal criteria guiding recruitment to these key positions. In this regard the recruitment of actors with capabilities to perform a leadership role is by no means ensured through institutional arrangements. Rather, the extent to which individuals with a potential to serve a leadership role are recruited into these positions is determined by governments.

7.4 Enhanced Effectiveness?

The analysis indicates that the provision of leadership functions has played a significant role in the IPCC process. Leadership behaviour seems to have served to facilitate both the development of the knowledge base (lead authors) and its transformation into premises for policy decisions (lead authors and especially WG chairs). It is particularly interesting to note, however, that the functions served by the WG and panel chairs both in the transformation of the knowledge base and in boundary roles seem to have had a significant impact on the nature of the effect of institutional arrangements and, hence, on the overall effectiveness of the process. The analysis thus suggests that the relationship between these two factors is characterised by contingency. The impact of institutional arrangements in terms of their capacity to enhance the effectiveness of the system depends in part on the provision of leadership. Conversely, the scope for and impact of

[9] The United States' chair of WGII and Canada's chair of WGIII after 1992 were, for instance, replaced without a formal vote in a plenary session.

leadership behaviour depends in part on institutional arrangements. Thus, while each of these factors clearly have had a significant and independent impact on the outcome of the process, the enhancement of the effectiveness of the IPCC process is above all linked to the manner in which these two factors complement each other.

This is particularly evident with regard to the functioning of one central institutional feature of the IPCC system; the submission of research results, assessments and summaries to critical review and the adversarial scrutiny by parties representing conflicting interests in the policymaking process. In chapter 6, we have emphasised this feature as one which has contributed significantly to the outcome of the IPCC process in the sense that it has had a direct impact upon the acceptability of the knowledge base to all "stakeholders". This chapter indicates, however, that the provision of leadership by agents operating in the system has contributed substantially to this effect, even to the extent that one may envisage a completely different (contrary) effect of this institutional arrangement had these leadership functions not been provided.

The emphasis on critical review and adversarial scrutiny inherent in the institutional design of the IPCC process implies that the system is vulnerable to politicisation and potentially paralysing conflicts. As discussed in chapter 6, the IPCC system is only to a little extent designed with a particular concern for the provision of institutional mechanisms for conflict resolution. Mechanisms for conflict resolution are subtle and based upon institutional arrangements designed for other purposes in the assessment process but which also may be utilised as instruments for handling disputes. While the institutional apparatus does provide tools and opportunities for handling disputes and resolving conflicts, their utilisation depends upon the behaviour of agents operating within the system. An important condition for the effectiveness of a design that serves to reveal conflict and bring them to the surface, therefore, is that instruments for conflict resolution provided by institutional arrangements are in fact utilised by agents operating in the system. Chapters 6 and 7 provide evidence of the effort made by the WG and panel chairs in this regard. Often facing a more or less constant threat of complete deadlock situations, the proceedings of both the WG and panel plenaries seem to have walked a thin line between success and failure. The end result, moreover, seems largely to have depended upon the substantive and personal capabilities of these actors to provide the necessary leadership. The provision of a knowledge base (assessments and summaries) under these conditions, whose factual validity and representativeness of the research field at large is accepted and acknowledged by both scientists and policymakers is a piece of scientific diplomacy well done. An entirely different outcome is not inconceivable, had these key actors not been capable of providing these leadership functions.

Another illustration of the contingent relationship between institutional arrangements and leadership behaviour is the differentiation of institutional arrangements according to function in the assessment process characterising the IPCC. The IPCC system is constituted by three institutionally separated decision-making levels with relatively few institutional links. This separation and the provision of institutional buffers between decision-making levels is one main device for the system's capacity to combine scientific integrity and political involvement in the process. The effectiveness of the process as a whole, however, is also dependent upon the provision of communicative links between the decision-making levels making up the system. When these communicative links cannot be provided by institutional means, the effectiveness of the system depends upon the ability of agents operating in the system to serve this function. While the role of the WG and panel chairs in the provision of this function is difficult to pinpoint in its precise nature, the chairs seem to have played a key role in the communication of the needs of policymakers and the limits of scientific advice to scientists and policymakers, respectively. In this regard, they have also been engaged in a continuous negotiation of the boundaries of science and policy. This analysis also suggests that the WG and panel chairs (in the panel plenary sessions) as well as lead authors (in WGI plenaries) have served an important communicative function in ensuring consistency between the various reports provided at the different decision-making levels: Lead authors have been instrumental in ensuring consistency between the assessments (provided by the scientific core) and their summaries (negotiated in WG plenaries). WG and panel chairs have been instrumental in ensuring consistency between the assessments (with summaries) of all the WGs and the synthesis report (negotiated in the panel plenary). In the development of a comprehensive problem diagnosis, this function seems crucial. Given the ever-present hidden agendas and governmental participants' constant awareness of possible policy implications, the provision of an inconsistent knowledge base could have been the outcome had this leadership function not been provided.

Finally, the contingency characterising the relationship between institutional arrangements and leadership behaviour is illustrated in the central role of the WG and panel chairs in the incremental development and negotiation of the institutional design itself. The formalisation of rules of procedure, carefully negotiated by the IPCC leadership, contributed to the development of a system in which governmental control at the upper decision-making levels is balanced by scientific control in the actual provision of assessments. These actors were also instrumental in the provision of a mandate for leadership for lead authors in WG plenaries. As we have seen, this (informal) device has become a key feature of the IPCC process in terms of both ensuring consistency between reports and, perhaps more importantly, in terms of maintaining the scientific authority of the knowledge base provided. It is also a key device for the provision of arenas for an interactive science–policy dialogue.

Above all, therefore, this analysis suggests that the impact of institutional arrangements depends in part upon the provision of several modes of leadership, and that the provision of leadership depends in part upon institutional arrangements. This implies that the effectiveness of processes of this kind is not determined and, hence, cannot be manipulated solely by design. An important determinant to the effectiveness of the process is the complementarity and the contingency characterising the relationship between these factors. How institutional arrangements work depends to some extent upon how agents operating within the system make them work.

This indicates that the shadow side of the effectiveness of institutional arrangements is the robustness of their impact. The relationship between effectiveness and robustness is complex. Their relationship seems, however, to be characterised by an element of inversion in the sense that effective institutional arrangements are also fragile, and that their impact is context-dependent and contingent upon other, less controllable factors such as the provision of leadership functions. The element of inversion characterising the relationship between the effectiveness of institutional arrangements and the robustness of this effect seems to be linked to the relationship between scientific integrity and some degree of science–policy integration. While it is possible to combine these institutional functions, their combination may have the undesirable side effect that the process simply does not work unless these vital leadership functions are also provided. In this regard, the inverse relationship between scientific autonomy and science–policy integration is in a sense reproduced in the element of inversion characterising the relationship between the (potential) effectiveness of institutional arrangements and the robustness of this effect. Thus, while our conclusion from chapter 6 that these dimensions may not be as hard to combine by institutional means as we expected is true, it is also true that their combination represents a challenge to agents operating within the system and that their capabilities of handling this challenge determines the extent to which their combination actually serves to enhance the effectiveness of the endeavour.

The provision of leadership depends not only on formal and documented qualifications, it also depends upon the personal capabilities, energy and skill of individual agents to convert a leadership potential into actual leadership. The options for ensuring the recruitment of individuals with the necessary capabilities to perform in leadership roles through formal recruitment criteria are therefore restricted. Moreover, in processes of this kind, the extent to which individuals with the necessary capabilities and skills to perform leadership functions are recruited into key positions may ultimately also depend upon the priorities and emphases of governments, which in turn, may be determined by their stakes in the process. Thus, the provision of leadership functions can by no means be ensured through institutional design. While institutional arrangements may be designed to facilitate the provision of leadership, idiosyncratic factors ultimately determine the extent of successful leadership provided. On the other hand, in a processes of this magnitude, where so many competent people participate, it is likely that some of them hold a potential to take on leadership roles. This analysis suggests that the outcome of the process may depend upon the extent to which these potential candidates are encouraged to do so.

Chapter 8

Causal Relationship: Real or Spurious?

8.1 Introduction

In chapter 5, we found that the science–policy dialogue taking place within the framework of the IPCC process seems to have been relatively effective. The extent to which policymakers have accepted the knowledge base provided is reflected in the development of a problem diagnosis, consensual among both scientists and policy makers. In the case of the IPCC, a knowledge base, whose representation of state-of-the-art knowledge in relevant fields is accepted and agreed upon by both scientists and policy makers, has indeed been developed and its linkage to policy goals, in terms of which societal activities that cause the problem as well as possible pathways for its resolution, have been identified and agreed upon. Thus, the IPCC process seems to have been effective in terms of levels one and two of the three-level definition of effectiveness developed in chapter 2.

The empirical analysis provided in chapters 6 and 7 suggests that the manner in which the IPCC process is organised and the extent to which leadership functions have been provided in the process are factors that have contributed substantially to this outcome. The aim of this chapter is to control, to the greatest extent possible, the robustness of this suggested causal relationship. In other words, could this outcome have been produced by other factors than the ones investigated in this analysis, thus implying that the causal relationship suggested by the empirical analysis is largely spurious?

Chapter 4 suggests two main alternative explanations to the outcome of the IPCC process: (i) explanations linked to the political setting of the problem, and (ii) explanations linked to the state of knowledge in the field.

The first set of alternative explanations may be represented by the following proposition:

> ***Alternative Proposition 1 (A1):*** *Policymakers' acceptance of the knowledge base observed in this case is a function of the configuration of political interests of parties (governments) in this problem area, and this outcome is only marginally affected by factors related to the science–policy dialogue.*

According to this line of argument, the level of effectiveness observed in this case is produced by factors related to the political setting of the problem rather than the features and functioning of the science–policy dialogue. The validity of this proposition may be investigated by analysing the malignancy of human-induced climate change as a political problem.

Chapter 4 defines political malignancy as the extent to which the problem has the features of a problem of incongruity, where actors under-represent actual costs and/or actual benefits in cost–benefit calculations upon which their actions (in relevant policy areas) are based. Such under-representation may be caused by at least two mechanisms: externalities and competition. To the extent that they are caused by competition, the malignancy of the problem is enhanced. Further, to the extent that the problem is characterised by asymmetries and/or cumulative cleavages, its malignancy is also enhanced.

To the extent that the climate change problem can be characterised as politically malign in these terms, proposition A1 is considered *not* to be supported: The more politically malign the climate change problem can be considered to be, the less likely it is that proposition A1 has factual validity.

The second set of alternative explanations may be represented by the following proposition:

> ***Alternative Proposition 2a (A2a):*** *Policymakers' acceptance of the knowledge base observed in this case is a function of the state of knowledge in this field of research, and is only marginally affected by factors related to the science–policy dialogue.*

This line of argument is linked to the assumption that scientific consensus leads to or facilitates the development of political consensus. This proposition, therefore, is linked to the extent to which state-of-the-art knowledge represented in the knowledge base provided is characterised by scientific consensus. While scientific consensus may be seen to encompass scientific uncertainty – in the sense that scientists agree on what they are uncertain about and hence where they may be in possible disagreement – scientific consensus is more often associated with relatively conclusive knowledge. In the following analysis, therefore, scientific consensus is seen as inversely related to scientific uncertainty.

The validity of proposition A2a is investigated in terms of the extent to which state-of-the-art knowledge represented in the knowledge base provided is characterised by scientific consensus. The conclusiveness of the conclusions is investigated in terms of the level of uncertainty with which they are associated. Proposition A2a is considered to be supported to the extent that the conclusions are associated with low levels of scientific uncertainty.

This proposition is also investigated in a slightly different formulation. Given that it is the IPCC itself that communicates the level of scientific uncertainty with which the knowledge base is associated, the policy acceptance we have observed in this case could have been produced by a misconception of uncertainty caused by a misrepresentation of uncertainty in the knowledge base provided by the IPCC. Hence, the impact of the state of knowledge is investigated not only in terms of its "objective" character, but also in terms of how it is represented in IPCC reports. This aspect, therefore, concerns the IPCC's handling and communication of scientific uncertainty, and may be represented by the following proposition:

> ***Alternative Proposition 2b (A2b):*** *Policymakers' acceptance of the knowledge base observed in this case is a function of their misconception of the state of knowledge caused by a misrepresentation of scientific uncertainty in the knowledge base developed by the IPCC, and is otherwise only marginally affected by factors related to the science–policy dialogue.*

This proposition is difficult to measure, but it is investigated in terms of the likelihood that parties may have misconceived the state of knowledge as a result of misrepresentations by the IPCC. We investigate the extent to which scientific uncertainty (and dispute) is addressed in IPCC reports, especially with regard to the extent to which there are points of dispute/uncertainty within the scientific community at large that are not well reflected in the IPCC reports. Obviously, we cannot investigate all areas covered by the IPCC reports. The validity of this proposition is, therefore, investigated in terms of examples: the IPCC's representation of the scientific debate concerning the sign of the water-vapour feedback effect. This proposition is also discussed in light of the debate that arose in the aftermath of the IPCC's conclusion that "there is a discernible human influence on global climate" in the SPM to the second assessment report.

Proposition A1 is investigated in section 8.2, and propositions A2a and A2b are investigated in section 8.3. In section 8.4 this discussion is summarised in terms of an assessment of the explanatory power of these factors as alternative explanations to the observed effectiveness in this case.

8.2 The Political Malignancy of the Problem of a Human-Induced Climate Change

The problem of a human-induced climate change bears the distinct characteristics of a collective action problem. A (global) climate system that is not dangerously affected by anthropogenic emissions of GHGs is an indivisible collective good. Once "provided", no one can be excluded from "consuming" it. Since catastrophic climate change has not yet occurred, and hence this good need not be provided, we may also see this problem in terms of the depletion of a common pool resource: The atmosphere is a common pool resource that is depleted by anthropogenic emissions of GHGs. In both cases, the maintenance of a climate system which is not dangerously affected by anthropogenic GHG emissions involves a free-rider problem: All rational actors (that are not altruists) want the good provided without having to bear any of the costs associated with its provision. Also, no actor can provide the good alone.

The nature of the climate change problem thus seems to correspond to Elster's definition of a "strong" collective action problem:

> According to the strong definition we have a collective action problem if two conditions are satisfied. First, each individual derives greater benefits under conditions of universal cooperation than he does under conditions of universal noncooperation. Second, each derives more benefits if he abstains from cooperation regardless of what the others do. (Elster, 1985: 139).

This definition implies that each actor has a dominant strategy of abstaining from co-operation. In game theoretic terms, this situation represents the "prisoner's dilemma". It is also possible, however, that at least some actors may adopt a strategy of co-operation on the condition that the number of other actors necessary to provide the good also adopt a co-operative strategy. If that is the case, it could modify the malignancy inherent in the situation. It is a highly unstable solution, however, since all (co-operating) actors would have incentives to abstain from further co-operation the moment the critical number of actors (C) is exceeded (C+1).

The climate change problem is, therefore, clearly not a problem of co-ordination. It has the characteristics of a problem of incongruity, where actors under-represent both the actual costs and the actual benefits in their cost–benefit calculations: The actual costs (of climate change) associated, particularly, with their energy policies (which is the dominant area of societal activities contributing to climate change), are not incorporated in their cost–benefit calculations. Hence, while they (may) share the costs (climate change) with all others, they reap the short-term benefits by themselves. Similarly, the benefits (no climate change) associated with a change in energy policies are not incorporated in their cost–benefit calculations.

This incongruity is mainly caused by competition. Competition clearly distorts actor incentives by amplifying the costs of co-operative behaviour. Unilateral regulations in the energy policy area may have significant negative implications for a country's international competitive situation since measures to stabilise or reduce GHG emissions to a large extent are associated with an increase in the price of energy. With a domestic increase in the price of energy that is not accompanied with similar increases in other countries, the competitiveness of the country that has unilaterally implemented GHG reduction measures is reduced. Second, if all countries implement GHG reduction measures, this is likely to have a significant impact on energy markets, in the sense that the demand for energy sources with the highest GHG emissions per energy unit (coal and oil) is likely to be reduced. Hence, even if the price remains the same or is increased, the total volume of the export is likely to be reduced and represent a loss to major exporting countries.

Finally, the climate problem is also characterised by asymmetries. While the regional impacts of climate change are (as we will also see below) highly uncertain, it is relatively clear that some regions will suffer more than others. Sub-Saharan Africa, for instance, will suffer relatively more than others because of its low capacity for adaptation. Similarly, small island states are likely to suffer more than others because of their vulnerability to sea-level rise. These regions also contribute very little to the problem itself. In this regard, the countries that "control" the problem in terms of their share of global GHG emissions are also the ones that are likely to suffer the least from it (see Underdal, 1997). Moreover, since such asymmetries also are associated with a country's adaptive capacity, which, *inter alia*, is linked to level of economic development, the climate change problem overlaps with the poverty problem, hence the North–South cleavage in international politics. In this regard, the conflicts generated by the climate problem are cumulative.

The climate change problem, therefore, bears all the characteristics of a malign political problem. Interests and preferences are highly divergent and conflicts are to some extent cumulative. The problem has asymmetrical elements in the sense that the actors who cause the problem are not likely to suffer most from it, and incentives to co-operate are distorted by elements of competition which serve to amplify the costs associated with a co-operative strategy. This complexity and malignancy is also reflected in the negotiations over GHG reduction measures.

Climate negotiations seem to a large extent to be characterised by a conflict structure with three main opposing parties: The United States, the European Community (EC) and the developing countries. *The United States* – a key player by virtue of its 22% share of global fossil emissions – was only very reluctantly dragged along to agree on the 1992 climate convention. The unwillingness of other OECD and developing countries to adopt a climate convention without the participation of the United States, implied that the United States to a large extent set the terms of the 1992 agreement (Nitze, 1994). The negotiating position of the United States succeeded in two important respects: it avoided a binding commitment on specified emissions reduction targets and timetables and won agreement on the "comprehensive approach" whereby mitigation strategies would encompass all GHGs[1] rather than just CO_2 (Nitze, 1994). The motivation for the refusal of the United States to adopt binding commitments on targets and timetables may be found partly in the traditional reluctance towards state interventionist policies. Instead they argued in favour of a flexible approach based on cost-effectiveness. The motivation may also partly be found in the ideologically based perception of a small group of presidential advisors led by former chief of staff John Sununu that the climate issue "was being used by environmentalists to impose their 'anti-growth agenda' on the US economy" (Nitze, 1994: 189; see also *Scientific American*, April, 1991). This perception made them unable to recognise the potential benefits the climate issue represented in terms of its potential linkage to increased investments in energy efficiency and renewable energy sources whereby the United States could enhance the cost-effectiveness of its own energy policies and reduce its oil import dependency (Nitze, 1994).

The Bush Administration pursued a hard bargaining strategy to prevent explicit reduction targets and timetables from entering the climate convention. President Bush even threatened to not attend the Rio meeting where the climate convention was supposed to be signed. A compromise was, however, achieved at the last minute after discussions with British environment minister Michael Howard, leading to the non-binding aim to stabilise emissions at their 1990 levels by 2000, agreed upon in article 4.2(b) of the convention (Nitze, 1994).

[1] Initially, the United States wanted "all GHGs" also to include Chlorofluorocarbons (CFCs) even though these gases were already regulated in the Montreal Protocol agreed upon in 1987. Their stabilisation target from 1990 of stabilisation of all GHGs at 1990 levels by 2000 was almost achieved the moment they made it because of their planned CFC phase-out in compliance with the Montreal Protocol (Paterson, 1992). The comprehensive approach adopted in the convention specifically exempts gases regulated by the Montreal Protocol from the GHG concept.

The negotiating position of the United States has been modified somewhat with the Clinton Administration. In 1993, Clinton announced American support for the target-oriented approach and declared that the United States was committed to the aim of stabilising GHG emissions at their 1990 levels by 2000 (Fermann, 1997). The internal opposition, particularly from the fossil fuels industrial lobby, against climate measures is still strong, however. Their main argument is that any climate measure, however extensive, made by the US government – with potentially dramatic consequences for the American economy – would be futile in the absence of emissions-reduction commitments from developing countries such as China and India. This position is supported by the Republican majority of the US Senate. Prior to the third meeting of the COP in Kyoto in December 1997, the Senate announced that an agreement that did not imply some sort of commitment from developing countries to curb their future GHG emissions would not be ratified. The pre-announced position of the United States at the Kyoto meeting was, therefore, a relatively unambitious goal of stabilisation of GHG emissions at 1990 levels in the period 2008–2012 (and hence actually a weaker obligation than its 1993 target). Moreover, there were three main conditions for its support of this target: that the convention allowed international quota trade and joint implementation; that the target encompassed all GHGs (not controlled by the Montreal Protocol); and, not least, developing-country commitments to curb future GHG emissions (Torvanger, 1997).

The European Community has, since the beginning, attempted to play a lead role in the drive for international regulatory policies to reduce GHG emissions. In 1990, the EC agreed upon a formal commitment to stabilise the CO_2 emissions of the Community as a whole at 1990 levels by 2000 (Paterson, 1992). Despite intense efforts to create a coherent EC policy to achieve this goal, mainly through a proposed EC carbon-energy tax, the EC experienced a setback in its ambitions during the period before the Rio meeting in 1992 (Wynne, 1993; Grubb, 1995). Largely due to strong industrial opposition and EC internal conflicts over the distribution of emissions reductions, the EC failed in its aspirations to develop a common position before the Rio meeting favouring a climate convention with explicit reduction targets and timetables (Wynne, 1993).

The efforts to develop a coherent EC strategy towards climate change took place during a period when the EC underwent major political and institutional changes with the development towards European integration and the 1992 Single European Market. The 1990 Maastricht Treaty, which implied a shift in policy sovereignty away from separate member states in several policy areas including environmental policies, generated some internal turbulence and controversy especially after the Danish referendum voted against it in 1992.

Being the only supranational entity with legally constituted policy sovereignty over its member countries, the EC is particularly well suited for concerted environmental action. National resistance against increased monetary and political integration has, however, to some extent been directed towards the development of environmental policies in the sense that some countries, for instance the UK, have targeted EC environmental policymaking as a particular candidate for repatriation (Wynne, 1993). Despite the uncertainties these controversies have generated, however, there has been a significant momentum within the EC towards concerted environmental policymaking. As pointed out by Wynne, "the EC member states share a common interest in creating a sufficiently united identity to be recognized as a global power in foreign policy, security and trade agreements in the new, post cold war world order" (1993: 102). The climate change negotiations represented a good opportunity for the EC to establish such a position.

The coincidence with the political integration process within the EC may have encouraged some countries "to guard their sovereignty jealously" in the development of a common EC strategy towards climate change (Wynne, 1993). In addition, the diversity among EC members with regard to level of economic development, the costs associated with different strategies and different policy instruments, resource availability, the role of fossil carbons in national economies, as well as national political cultures and approaches to policy making are significant (see, for instance, Wynne, 1993; Grubb, 1995). The development of a common strategy would, therefore, require a substantial amount of EC internal negotiations. Until as recently as 1997, the extent to which one could speak of an "EC climate policy" at all has been questioned. For instance, the 1990 stabilisation target has been suggested to amount to no more than the sum of each member countries' unilateral commitments (Wynne, 1993; Grubb, 1995). Moreover, as the Kyoto meeting approached rapidly, the extent to which the EC would manage a stabilisation target at all was increasingly questioned[2] (Grubb, 1995).

Therefore, it came as a great surprise to most close observers when the member states of the EC, in March 1997, agreed to a negotiating position to reduce total EC emissions of the three major GHGs (CO_2, CH_4, and N_2O) by 15% in 2010 compared to 1990 levels, and proposed that OECD countries cut their GHG emissions by an equal percentage (Ringius, 1997). This became the toughest policy target proposed by industrialised countries in the

[2] It should also be noted, however, that the fuel switch from coal to gas implemented in the UK independently of the climate issue could amount to a 10% reduction in British CO_2 emissions (Ringius, 1997).

Kyoto protocol negotiations. Furthermore, the target was to be met through a complex differentiation of commitments among EC countries, where countries like Germany and Denmark were committed to 25% GHG emissions reductions while, for instance, Portugal was allocated a 40% increase[3] (Ringius, 1997).

Burden-sharing is essentially a zero-sum game. Given that a group of countries intends to reduce a given total amount of emissions, the less GHG emissions one country reduces, the more GHG emissions others will have to reduce (Ringius, 1997). Ever since the EC became involved in climate negotiations in the early 1990s, EC member countries realised that some sort of burden-sharing formula had to be agreed upon within the EC in the development of a common position. However, throughout the seven years of negotiations (since the 1990 EC stabilisation target), they had not succeeded in developing such a burden-sharing formula. From 1995 (the first meeting of the COP), efforts to develop a common position at the Kyoto meeting implied intensified negotiations over internal burden-sharing. Until the Dutch Presidency period, which commenced 1 January 1997, prospects for success seemed meagre. The Dutch government was, however, well prepared for the task, and had engaged a team of experts from the University of Utrecht for assistance. The "triptique approach" was the result, with Bert Metz, the chief Dutch government participant at the climate negotiations, as its primary architect (Ringius, 1997). In contrast to the earlier efforts to develop burden-sharing formulas within the EC – which represented "vertical" approaches, mainly focused on each countries' measures and policies and according to which each member country was treated as a "black box" – the triptique approach represented a "horizontal" approach, which focused rather on economic sectors across the EC. Hence, national economies were separated in three broad sectors: the light domestic sector; the energy-extensive, export-oriented sector; and the electricity generation sector. Taking the existing sector structure as a reference point, burden-sharing was achieved by adopting identical "targets" or "allowances" for each sector across the EC, in accordance with some criteria agreed upon among member countries through a complex and difficult negotiation process.[4] Thus, by developing a cross-sectoral approach, the EC managed to agree upon a burden-sharing formula that facilitated agreement on a common target that implied substantial emissions reductions and enabled them to maintain their leadership role in climate negotiations.

Developing countries have a substantial influence potential in global negotiations by virtue of their sheer size. The group of developing countries

[3] After the Kyoto agreement, the differentiation of commitments within the EC has been renegotiated.

[4] For a discussion of the development of the triptique approach, see Ringius, 1997.

is, however, at least as heterogeneous – with regard both to energy situation and negotiating positions – as the industrialised countries. They have thus experienced problems with co-ordinating their positions. The group of developing countries comprises the Group of 77 and China. In addition, the group of small island states – which formed an alliance during the early phase of the negotiations (AOSIS) – sometimes participates with the Group of 77.

Among developing countries there are at least three major sub-groups: A first sub-group is constituted by the biggest developing countries, principally China, India and Brazil, whose influence derives from the fact that their proportion of global GHG emissions is expected to rise dramatically during the next few decades. Scenarios suggest that the current 26% contribution to global fossil CO_2 emissions from developing countries will rise to more than 50% by 2025 (Paterson, 1992). A large proportion of this increase will come in these countries. In a business-as-usual scenario, China will become the world's largest emitter of CO_2 in 2020 (Beuermann, 1997). Moreover, the huge abatement potential in these countries is illustrated in that in China alone, the possible reduction of CO_2 in 2020 equals the 1990 energy-related CO_2 emissions of the UK (Beuermann, 1997). This situation gives these developing countries considerable leverage, since their role in a climate regime will have a crucial impact on the ultimate effectiveness of the regime. Their position has been centred around an uncompromising requirement that all developing country commitments are to be conditional on provision of technology and finance from industrialised countries (Paterson, 1992; Dasgupta, 1994).

A second sub-group comprises the major oil-producing developing countries. This group has been led by Saudi Arabia and Kuwait, but has included also other Middle Eastern oil producers, as well as Algeria, Venezuela and, to some extent, Nigeria (Paterson, 1992). For obvious reasons, this group is concerned about the impacts commitments by industrialised countries to reduce fossil-fuel-related emissions may have on the petroleum market and hence their national economy. Consequently, this group has opposed suggestions that industrialised countries should set targets to reduce fossil-fuels-related GHG emissions. On some occasions, some members of this group have taken a position similar to that of the United States (Dasgupta, 1994).

A third sub-group comprises the least developed countries, largely Sub-Saharan Africa. In addition to the general focus of developing countries on the need for technology and financial transfers, this group has also focused more strongly on their general vulnerability to climate change and abnormal weather conditions – a change in precipitation patterns towards a drier climate in particular – and their needs in relation to adapting to climate change. This group currently contributes very little to global GHG emissions, and is unlikely to do so in the immediate future. The influence of this group is, therefore, dependent upon general support from the Group of 77.

The alliance of small island states, AOSIS, has functioned both as a participant of the Group of 77 and as an autonomous coalition. Their primary concern is the potentially dramatic consequences they may experience from a sea-level rise induced by global warming. While this alliance is small both in number and, especially, in terms of their share in global GHG emissions, they have reportedly been extremely successful at getting their message across in the negotiations (Paterson, 1992). Represented by "able British and American lawyers and academicians", this coalition has vigorously argued the case for a strong climate convention (Dasgupta, 1994).

Given these widely diverging interests, developing countries have obviously experienced problems with co-ordinating their negotiating positions. In general, however, developing countries have sought acknowledgement from industrialised countries that the primary responsibility for causing the climate problem, and hence, the primary responsibility for taking measures for solving it lies with the industrialised countries. Moreover, they managed to develop a common position on the section of the convention concerning "principles". Further, they agreed that financial transfers should be "adequate, new and additional resources" and technology transfers should be made on "favorable concessionary and preferential terms" (Dasgupta, 1994). They also agreed on formulations to the effect that any commitment by developing-country parties be conditional upon an obligation of developed-country parties to provide financial and technology transfer (Dasgupta, 1994). Given that the positions of AOSIS and the oil-producing developing countries represented opposite poles in the negotiations, they did not succeed in their efforts to develop a comprehensive consensus text covering the Group's position on commitments (Dasgupta, 1994).

Developing countries have thus agreed upon a basic position made up of two main elements: (i) To the extent that GHG emissions are to be reduced (to which the oil-producing developing countries are opposed), it is developed countries that must bear the responsibility for such reductions; and (ii) To the extent that developing countries do take on commitments in the climate convention, they must be conditional upon additional financial and technology transfers.

The industrialised countries were unwilling, in clear language, to acknowledge their responsibility for the climate problem and its solution. Hence, a Chinese proposal to insert text to this effect in the preamble of the convention did not succeed (Dasgupta, 1994). The climate convention does, however, contain more or less ambiguous language concerning the "specific needs and special circumstances of developing country Parties" (article 3.2. See also articles 3.4 and 3.5). Initially, developed countries were also hesitant to accept a formulation concerning financial commitments of "new and additional" resources. Disagreement prevailed between developed and developing countries over whether financial resources should be provided on a voluntary or assessed basis and whether they were intended to meet the "full" or the "agreed" incremental costs of developing countries (Dasgupta, 1994). In the convention it is recognised, however, that

> [t]he extent to which developing country Parties will effectively implement their commitments under the Convention will depend on the effective implementation by developed country Parties of their commitments under the Convention related to financial resources and transfer of technology and will take fully into account that economic and social development of poverty eradication are the first and overriding priorities of the developing country Parties. (Article 4.7)

The parties also agreed to a compromise text which states that the "agreed full incremental costs" should be provided by developed countries through provision of "new and additional financial resources" (Article 4.3) (Dasgupta, 1994). As of yet, developing countries have refused to agree to any commitments to curb future GHG emissions.

In sum, therefore, climate negotiations are, first, characterised by significant controversies within as well as between the North and the South. Hence, each of these blocs have spent considerable time and energy on internal negotiations. During the negotiations over the climate convention, these blocs were so preoccupied with internal negotiations that they barely had the time to negotiate with each other. As pointed out by Dasgupta, right up until the fifth and final session of the INC, "there were few attempts at bargaining or negotiating to narrow down the wide substantive differences between the North and the South. Indeed negotiations did take place within each group aimed at formulating common positions but there was little effective bargaining between North and South" (1994: 140–1).

Second, the conflicts both within and between these blocs were characterised by incompatible positions. For instance, industrialised countries battled with difficult conflicts over whether or not commitments were to be differentiated, and, if they were, according to which criteria differentiated commitments were to be allocated. This question, therefore, did have zero-sum characteristics: Either commitments were differentiated or they were not. If they were to be differentiated, moreover, given that the total amount of GHG reductions was fixed, the bargaining problem of the differentiation itself took on a zero-sum nature. It should also be noted that these conflicts prevailed between two parties that are powerful both in the basic game and in the negotiation game: namely, the EC and the United States (see, for instance, Underdal, 1997: 9). The developing countries fought similar battles particularly between the AOSIS, which proposed that developed countries should reduce their CO_2 emissions by 20%, and the oil-producing developing countries, who were opposed to all emissions reduction targets. A similar, but perhaps less intense, conflict also characterised the relationship between the biggest developing countries – China, India and Brazil – and the oil-producing developing countries. China, India and Brazil supported a commitment by industrialised countries to reduce their GHG emissions, but they were principally concerned about safeguarding their own position in a future climate regime without commitments to curb their own GHG emissions.

Third, the nature of this problem is characterised by a mismatch between the countries that may be assumed to be most influential, both in the basic game and in the negotiation game – the USA, the EC, China, India and Brazil – and the countries that may be assumed to be most vulnerable to climate change, China, AOSIS and Sub-Saharan Africa (see, for instance, Underdal, 1997: 9). Hence, the countries that "control" the problem (by virtue of the size of their GHG emissions) are to a large extent the ones that are likely to suffer the least from it (with the exception of China).

Finally, the question of whether or not developing countries such as China, India and Brazil should make commitments to curb their future GHG emissions became a major point of controversy between the North and the South during the Kyoto meeting. With the announced position of the US Senate in their luggage – that a protocol which did not imply commitments by these developing countries would not be ratified – the US delegation negotiated intensely to this effect at the Kyoto meeting. When the US delegation finally did accept a protocol where they committed themselves to a 7% reduction in GHG emissions without any concessions from these developing countries after all, a great deal of the explanation may be found in the opportunities the protocol provides for international quota trading and a semi-official agreement between the United States, Canada, Japan and Russia. Fred Pearce reports that only hours before the Kyoto summit was due to end, "the US announced the formation of a 'carbon club' consisting of itself, Canada, Japan and Russia, to trade emissions permits among themselves" (*New Scientist*, 20/27 December 1997). Russia's current CO_2 emissions lie 30% below their 1990 levels, and are unlikely to recover more than two-thirds of that ground before 2008, when the trading will begin. With this agreement in place, the United States agreed to raise its target to 7%. The Europeans were presented with a *fait accompli*: Either to accept this deal, or to wreck the conference. They accepted, even though energy analysts advised that trading would permit the United States and Japan to raise their domestic emissions 12–16% above national targets (*New Scientist*, 20/27 December, 1997).

The political setting of the climate change problem, therefore, is characterised by a significant level of political malignancy. This implies that it is highly unlikely that parties have (opportunistically) accepted the knowledge base provided by the IPCC as legitimisation for the realisation of interests they would have pursued anyway in the concerned policy areas. First, for some parties, the acceptance of the climate change problem causes more problems and costs than new opportunities and benefits. Second, the opportunities that are provided by this problem are also often conditional upon concerted action. Third, even among relatively homogenous countries such as, for instance, OECD member countries, the opportunities and costs associated with energy policies that would serve to reduce emissions of GHGs are distributed very unequally. The political setting of the problem, therefore, cannot explain the level of policy acceptance of the knowledge base provided by the IPCC indicated by the policy response to this problem. On these grounds, it is unlikely that factors associated with the nature of the political setting – regardless of factors related to the science–policy dialogue – have produced the level of effectiveness observed in this case, and that alternative proposition A1 is unlikely to have factual validity.

8.3 The State of Knowledge

The second alternative proposition concerns the extent to which policy makers' acceptance of the knowledge base provided by the IPCC may be explained by the state of knowledge in this field of research, and hence, would have come about regardless of factors related to the nature and functioning of the science–policy dialogue. This proposition concerns the relationship between scientific consensus and political consensus, and is based upon the assumption that the former leads to, or at least facilitates, the latter. While this assumption has been questioned (see, for instance, Jasanoff, 1990), there is also evidence that it has some validity (Andresen et al., in press). The second alternative proposition, therefore, is based on the line of argument that the state-of-the-art knowledge represented in the knowledge base provided by the IPCC is characterised by a conclusiveness which may have led to the level of acceptance observed, regardless of factors related to the process whereby it was communicated.

8.3.1 The Scientific Uncertainty of Climate Change

A key element in the debate about the size and nature of projected climate changes is the credibility of the mathematically and physically based climate models that are used to project the climate change resulting from an increased accumulation of GHGs in the atmosphere caused by human activities. Mahlman's[5] "policy-independent" evaluation (his own formulation) of the levels of current scientific confidence in predictions emanating from climate models is useful in assessing the nature and level of scientific uncertainty characterising this field of research (Mahlman, 1996).

Mahlman distinguishes between *virtually certain 'facts'*; *virtually certain projections* (greater than 99 out of 100 chance of being true within the predicted range); *very probable projections* (greater than 9 out of 10 chance of being true within the predicted range); and *probable projections* (greater than 2 out of 3 chance of being true). The first category includes key aspects in the knowledge about the climate system that do not depend directly upon the performance of climate model simulations. The following propositions may, according to Mahlman, be stated as "virtually certain 'facts'":

- That atmospheric concentrations of GHGs are increasing because of human activities.
- That the altered amounts of GHGs will affect the climate for many centuries.
- That human-caused increases of CO_2 and decreases of ozone in the stratosphere have produced a stratospheric cooling which is consistent with model predictions[6] (more than 1°C average cooling).
- That the Earth's surface has warmed by about 0.5°C (± 0.2°C) over the past century.

According to Mahlman, it is also a fact, however, that, "the uncertainties concerning the responses of clouds, water vapor, ice, ocean currents, and specific regions to increased greenhouse gases remain formidable" (Mahlman, 1997: 1416).

The following propositions may, according to Mahlman, be regarded as "virtually certain projections":

- That the stratosphere will continue to cool significantly as CO_2 increases and that if ozone also continues to decrease, the cooling will be magnified.

[5] J. D. Mahlman is at the Geophysical Fluid Dynamics Laboratory, NOAA, Princeton University, and has not participated in the IPCC process except as a reviewer.

[6] Stratospheric cooling is considered to be part of a "fingerprint" that attributes global warming trends to human activities (see, for instance, Nicholls, 1996).

– That global mean amounts of water vapour (the most important GHG) will increase in the lower troposphere by roughly 6% per 1°C of warming.

All the main conclusions regarding the response of the climate system to the increased levels of CO_2 in the atmosphere may, according to Mahlman, be considered "very probable projections":

– That the global warming observed over the past century is generally consistent with model predictions, if a reasonable sulphate particle offset is included, and that it is difficult to construct conceivable alternative hypotheses to explain this observed warming.
– That a doubling of atmospheric CO_2 over pre-industrial levels is projected to lead to an equilibrium global warming in the range of 1.5° to 4.5°C. This generous range also indicates the levels of uncertainty, which are primarily linked to the radiative feedback effects of clouds and water vapour in the upper troposphere.
– That there is a roughly 10% chance that the actual equilibrium warming caused by doubled CO_2 levels could fall outside this range (be lower than 1.5° or higher than 4.5°C).
– That the projected sea level rise of 50 ± 25 cm by 2100 is below the equilibrium sea level rise that would ultimately be expected.
– That global mean precipitation will increase by about $2 \pm 0.5\%$ per 1°C warming.

The propositions that may be considered probable projections concern the more detailed accounts of the response of the climate system to a sustained increase of atmospheric GHGs and are not discussed further here.

Mahlman's account of the confidence scientists have in climate model simulations and projections thus indicates that while scientists consider it virtually certain that the anthropogenic increases in atmospheric GHG concentrations will affect the climate for many centuries, they are uncertain about the precise nature of the response of the climate system, reflected, for instance, in the broad range of temperature responses to be expected from a doubling of atmospheric CO_2. They are also uncertain about when these responses may occur – an uncertainty which, of course, also is linked to uncertainties in future emission trends.[7] The rate of change is particularly important since the adaptive capacity of both human societies and bio-systems depends on how rapidly the changes take place. Even large climate

[7] In his evaluation, Mahlman has used two extrapolations of the IPCC scenarios of increases of 1% equivalent atmospheric CO_2 concentrations per year: The first case levels off at a CO_2 doubling after 70 years, the second levels off at a CO_2 quadrupling after 140 years. Mahlman emphasises that "these leveled-off scenarios are physically plausible but are presented as illustrations, not as social predictions" (1997: 1416).

changes may be undramatic as long as they take place over sufficiently long periods of time, permitting human societies and bio-systems to adapt. Finally, the attribution of the warming trend observed to anthropogenic increases in the atmospheric concentrations of GHGs is associated with significant uncertainty due to the natural variability of the climate system. Mahlman's evaluation, therefore, gives us a clear impression of the amount of scientific uncertainty with which this field of research is associated.

The knowledge base provided by the IPCC is associated with major points of uncertainty that may not be resolved for quite some time. Moreover, the improvement in the state of knowledge which has taken place during the course of the IPCC process has revealed (new) scientific uncertainty as much as it has served to reduce scientific uncertainty. Thus, although there has been a marked improvement in the state of knowledge during the course of the IPCC process, it is not comparable, for instance, with the improvement in knowledge that took place in the ozone process. The improvement in the state of knowledge that took place during the course of the ozone process implied that policymakers were confronted with increasingly more compelling evidence that emissions of CFCs and other substances were in fact depleting the ozone layer (see, for instance, Benedick, 1990). This has not occurred in the process on a human-induced climate change. Indeed, the level of scientific uncertainty associated with the climate change problem is one main argument for opposing political action in this policy area.

In this case, policymakers have thus been confronted by a political problem that is characterised by significant levels of scientific uncertainty. This also has significant implications for their decision situation. The uncertainty concerns elements that are highly relevant for the development of an appropriate policy response such as regional variations, the timing of the climate system's response to enhanced concentrations of atmospheric GHGs, and its magnitude. It thus seems unlikely that the state of knowledge has produced the policy response in this area and that factors related to the effectiveness of the science–policy dialogue only have had marginal impact on the outcome. Rather, this level of uncertainty is more likely to have implied that the effectiveness of the dialogue has increased in importance. It is precisely under conditions of uncertainty that information attains a crucial strategic value to policymakers both as a guide to action and as a tool for realising political interests. Thus, uncertainty increases the parties' demand for reliable information and knowledge while it also reinforces incentives to information distortion and manipulation. The capacity of the institutional apparatus within which the science–policy dialogue takes place to cope with the pitfalls inherent in science–policy interaction may thus be even more important for the process to succeed in a situation characterised by this level of uncertainty.

These mechanisms also suggest, moreover, that there may be significant interaction effects between the state of knowledge and political malignancy. In the case of climate change, for instance, scientific uncertainty implies that the task of quantifying damage effects of climate change and, hence, potential benefits resulting from abatement strategies is very difficult, if not impossible. Uncertainty, therefore, reinforces an actor's incentives to misrepresent the true costs and benefits associated with their actions. On this basis, therefore, proposition A2a is considered unlikely to have factual validity.

8.3.2 The IPCC's handling of Scientific Uncertainty

The IPCC's handling of this uncertainty has been a central element in the public debate over a human-induced climate change. It has, for instance, been suggested that the IPCC has under-stated *and* over-stated the level of scientific uncertainty in different phases of the process as an element in a strategy to generate more research funds (under-statement during the initial phase of the process to get the issue onto the international political agenda, and over-statement during the later phases of the process to maintain the financial flow to the research community) (Boehmer-Christiansen, 1994; for a comment on this view, see also Shackley and Skodvin, 1995). The question remains, therefore, whether the IPCC has manipulated the level of conclusiveness characterising the knowledge base and thereby led policy makers to believe that the knowledge base is more conclusive than it actually is. If that is the case, policy makers may have responded to (their perception of) the conclusiveness of the knowledge base regardless of factors related to the process whereby it was communicated – except for the IPCC's eventual manipulation.

The IPCC's handling of scientific uncertainty is discussed in relation to two particular cases: the IPCC's representation of the scientific debate concerning the sign of the water-vapour feedback effect, and the debate that arose in the aftermath of the IPCC's conclusion that "there is a discernible human influence on global climate" in the SPM to the second assessment report. Both these cases are related to key elements and key uncertainties in climate change research and are therefore seen as representative of the IPCC's handling of scientific uncertainty in general.

The Water-Vapour Feedback Effect

Interactive feedback mechanisms constitute the greatest uncertainty in projections of future climate changes associated with increased atmospheric concentrations of GHGs (Cess, 1991). In particular, the sign of the feedback effect of water vapour has been debated. Given that water vapour is the most important GHG in the atmosphere, the sign of its feedback – whether an increase in temperature is associated with an increase or a decrease in water vapour – is a key element in the theory of a human-induced climate change. All numerical models calculate a positive feedback effect of water vapour. If this feedback effect is negative, it would have very significant implications for the total warming effect to be expected from anthropogenic increases in atmospheric levels of GHGs, since this (cooling) effect would be strong enough to almost counterbalance the entire warming effect caused by anthropogenic GHG emissions. In this regard, the theory of a human-induced climate change depends heavily on a positive feedback effect from water vapour.

Dr. Richard Lindzen – one of the main critics of the IPCC process in general, and the scientific conclusions brought forward in IPCC reports in particular – maintains that all General Circulation Models (GCMs) erroneously parameterise a positive water-vapour feedback effect (Lindzen 1990a). While he agrees that a warmer climate would lead to an increase in water vapour near the ground, he argues that a warmer climate is also associated with "more and deeper cumulus convection", which leads to a drying of the upper troposphere (Lindzen, 1990a: 296). It is, according to Lindzen, crucial to understand the special role of water vapour in the upper troposphere since, "it is at these levels that the trapping of heat by greenhouse gases assumes its greatest importance" (Lindzen, 1990b: 1465). Thus, Lindzen argues that even though the feedback of water vapour in the lower troposphere is positive, the drying of the upper troposphere he claims to be associated with a warmer climate could lead to an overall negative water-vapour feedback effect:

> It is easily shown that, on a molecule for molecule basis, water vapor above 6 km is much more important than water vapor near the ground in determining temperature (…). Thus, even though water vapor near the ground increases with increasing temperature, a smaller absolute decrease at upper levels can lead to a negative feedback. (1990b: 1466)

It should perhaps be mentioned that while Lindzen argues that this possibility is there, he does not demonstrate the actual performance of this mechanism:

> The above hardly exhausts the possible sources of error in present models. However, it suffices to show that the *possibility* of large overestimates exists. Consistent with past data [Lindzen also claims that there is no observational evidence of a warming trend], corrected models may very well end up predicting greenhouse warmings of only a few

tenths of a degree centigrade. Such changes have already occurred (...) without disastrous consequences. (Lindzen, 1990a: 297, emphasis in original)

How then, has the question of the sign of the water-vapour feedback been dealt with by the IPCC? IPCC scientists undoubtedly under-estimated the complexity of water-vapour feedback in the First Assessment report, where it is stated that:

> [t]he best understood feedback mechanism is water-vapour feedback, and this is intuitively easy to comprehend. ... an increase in one greenhouse gas (CO_2) induces an increase in yet another greenhouse gas (water vapour), resulting in a positive (amplifying) feedback mechanism. ... The point is that water-vapour feedback has amplified the initial global warming of 1.2°C to 1.9°C, i.e., an amplification factor of 1.6. (Cubasch and Cess, 1990: 78)

In the Executive Summary of the report it is even stated with "certainty" that "the main greenhouse gas, water vapour, will increase in response to global warming and further enhance it" (IPCC, 1990: xi). This statement could perhaps qualify as a gross exaggeration of the state of knowledge at that time, given the (then) recent publication of Lindzen's paper.

The IPCC's First Assessment report was accepted by the IPCC plenary in June 1990. Lindzen's paper was published in March 1990. Thus, Lindzen's approach to the question of water-vapour feedback could have been discussed, or at least mentioned, in the IPCC report, especially given the potentially significant implications a negative water-vapour feedback would have for calculations of a human-induced climate change. Thus, while the chapter was probably written in 1989 and hence before Lindzen's publication, the failure to include and discuss Lindzen's argument in later revisions of the report must be regarded as an omission and an under-estimation of the complexities associated with this question.

After the First Assessment Report, a general criticism was raised towards WGI for a failure to discuss dissenting views. In an editorial in Nature, for instance, it is stated that

> IPCC's failure to discuss dissenting opinions, perhaps to dismiss them, was a mistake. If, for example, one such as Dr Freeman Dyson should go about the world (as he does) saying that the disappearance of half the carbon dioxide added each year to the atmosphere casts doubt on the standard picture of global warming, he deserves and answer. (Nature, vol. 348, 15 November 1990, pp. 182).

While the controversy implied in Lindzen's paper over the sign of the water-vapour feedback is not given attention in the first IPCC assessment, a general discussion of the scientific uncertainty with which the research findings are associated is provided in the report; there is a separate chapter on scientific uncertainty and how to deal with it (McBean and McCarthy, 1990), and scientific uncertainty is also commented upon both in the Executive Summary and in the Summary for Policymakers. In the executive summary, for instance, it is stated that "there are many uncertainties in our predictions particularly with regard to the timing, magnitude and regional patterns of climate change" (p. xii). Nevertheless, the formulations of the executive summary – particularly the headings: *we are certain of the following...* and *we predict with confidence that...* – made good headlines in the media. These formulations may have led, if not policymakers – who can be expected to read the summaries in their entirety – then at least the public to believe that the scientific controversies over an enhanced greenhouse effect were more settled than they actually were.

During their work with the 1992 Supplementary Report, the leadership of WGI made several efforts to persuade Lindzen to come to the lead- and contributing-author meetings where the water-vapour feedback effect was to be discussed. Lindzen never showed up, but he did send a few pages on the issue for the WG to consider (personal communication with WGI leadership). In the supplementary report, the uncertainties surrounding upper tropospheric water vapour are much more extensively discussed than in the first assessment. In the SPM, for instance, it is stated that "there has been some clarification of the nature of water-vapour feedback, although the radiative effects of clouds and related processes continue to be the major source of uncertainty and there remain uncertainties in the predicted changes in upper tropospheric water vapour in the tropics" (Houghton et al., 1992: 15). The question of the sign of water-vapour feedback is discussed at length in the report itself, and Lindzen's arguments in his 1990 publication are specifically discussed (Gates et al., 1992: 115). Thus, the main conclusion – which still suggests a positive water-vapour feedback – is better documented than in the first assessment report: "There is no compelling evidence that water-vapour feedback is anything other than the positive feedback it has generally been considered to be, although there may be difficulties with the treatments of upper-level water vapour in current models" (Gates et al., 1992:114).

In the 1992 Supplementary Report, Lindzen is listed both as contributing author and reviewer. This has been commented by Lindzen:

> [The] statement that all listed scientists participate in the decisions is simply untrue. That a scientist may be invited to write a section on short notice (under 3 weeks) does not mean that the IPCC will take any cognizance of what is written, or that the scientist will ever be consulted. He is nonetheless listed as a participant. In my case, I was not consulted

> on the final or even preliminary versions of anything. A participant can increase his impact by attending a series of meetings around the globe, but this is not a viable option for most of us who teach and have limited time and funds. (Lindzen, 1993)

Thus, Lindzen is strongly critical of the nature of the IPCC process: "I wouldn't touch the IPCC because it is an illegitimate process" (personal communication).

The extensive discussion of the water-vapour feedback is continued also in the second assessment report, including a specific mentioning of Lindzen's argument (Dickinson et al., 1995: 200). An extensive discussion on the nature of the water-vapour feedback, including a discussion of publications in which Lindzen's theory has been tested against empirical observations and calculations, is concluded with the following statement:

> Feedback from the redistribution of water vapour remains a substantial uncertainty in climate models. That from the lower troposphere seems least controversial. Much of the current debate has been addressing feedback from the tropical upper troposphere, where the feedback appears likely to be positive. However, this is not yet convincingly established; much further evaluation of climate models with regard to observed processes is needed. (Dickinson et al., 1995:201)

The IPCC's initial handling of the arguments presented by Lindzen concerning the sign of the water-vapour feedback may suggest that IPCC scientists grossly under-estimated the scientific uncertainty with which state-of-the-art climate research was associated. This under-statement of scientific uncertainty concerning a major point during the initial phase of the IPCC process has been maintained to represent an element in a conscious strategy by the IPCC leadership to generate more research funds by getting the issue on the international political agenda (Boehmer-Christiansen, 1994). While the failure to discuss this point seriously may be reprehensible, it is difficult to believe that this constituted a conscious strategy by the leadership of the IPCC as suggested by Boehmer-Christiansen. Given the very nature of the IPCC process, with its intense scientific and political scrutiny of findings and conclusions as discussed in the empirical analysis here, there is no evidence suggesting that the IPCC leadership had this kind of influence over either the reports or, particularly, the summaries. If this were a conscious fund-generating strategy, it would have included a large number of scientists – indeed, almost the entire scientific community. No evidence that an entire community of scientists followed or was led into this strategy of deception has been found. Rather, the evidence suggests that a more credible and likely interpretation is that IPCC scientists actually believed that matters were more settled than they turned out to be. Thus, this analysis suggests that the understatement concerning the scientific controversy about the water-vapour feedback is not an indication of a conscious manipulation by IPCC scientists, but rather in itself a consequence of scientific ignorance. Even though this neglect of the uncertainty associated with this factor is not the result of a conscious strategy of deception, it may have caused a misguided perception among policymakers of the conclusiveness of the knowledge base. It should, however, be emphasised that major points of uncertainty were communicated also in the first assessment report, even though this particular point was neglected.

The Controversy over "A Discernible Human Influence on Global Climate"

The strongest criticism of the IPCC's account of scientific uncertainties – which perhaps received most public attention – arose in the aftermath of the WGI plenary meeting in Madrid in 1995, where the WG's contribution to the second assessment report and its accompanying SPM were accepted and approved. In the SPM the WGI plenary agreed, after considerable debate over three days, that, "the balance of evidence suggests that there is a discernible human influence on global climate" (p. 11). As discussed above (chapter 6, section 6.3.2, footnote 8), the Madrid meeting did not have enough time to finalise their revision of the document originally intended to be the SPM. Hence, the executive summary was converted into an SPM, and the document originally intended as the SPM was converted into a "technical summary", which was to be submitted to the full panel plenary for acceptance. The short time allocated for the WGI plenary meeting implied that the incorporation of comments from the review process, as well as from the WGI plenary meeting itself, into both the assessment report and the technical summary was not finalised when the meeting ended. For this reason, editorial changes – of a much more substantial nature than the usual final editing of IPCC reports – were made to the assessment report after the WGI meeting. These changes became subject to a heated public debate in the aftermath of the meeting and, particularly, in the aftermath of the report's publication. The changes made to the final paragraph of chapter 8 (the chapter dealing with the most controversial issues of detection and attribution of climate change) are representative of the changes made, and are therefore cited at some length.

In the *draft* report, the final paragraph reads:

> Finally, we come to the most difficult question of all: 'When will the detection and unambiguous attribution of human-induced climate change occur?' In the light of the very large signal and noise uncertainties discussed in this Chapter, it is not surprising that the best answer to this question is, 'We do not know.' Some would and have claimed, on the basis of the results presented in Section 8.4, that detection of a significant

> climate change has already occurred. Few if any would be willing to argue that unambiguous *attribution* of this change to anthropogenic effects has already occurred, or was likely to happen the next several years.[end of paragraph] (Draft report of 9 October, 1995, Chapter 8, p. 8.17, emphasis in original).

In the *final* report, the corresponding paragraph has been enlarged and rewritten, and reads:

> Finally, we come to the difficult question of when the detection and attribution of human-induced climate change is likely to occur. The answer to this question must be subjective, particularly in the light of the large signal and noise uncertainties discussed in this chapter. Some scientists maintain that these uncertainties currently preclude any answer to the question posed above. Other scientists would and have claimed, on the basis of the statistical results presented in Section 8.4, that confident detection of a significant anthropogenic climate change has already occurred.
>
> As noted in Section 8.1, attribution involves statistical testing of alternative explanations for a detected observed change, and few would be willing to argue that *completely unambiguous* attribution of (all or part of) this change has already occurred. However, evidence from the pattern-based studies reported on here suggests that an initial step has now been taken in the direction of attribution, ...
>
> The body of statistical evidence in Chapter 8, when examined in the context of our physical understanding of the climate system, now points towards a discernible human influence on global climate" (Santer et al, 1995:439, emphasis in original).

The re-editing of the second assessment was severely criticised by the Global Climate Coalition (GCC), which accused the IPCC of "scientific cleansing" (*Nature*, vol. 381, 13 June 1996; see also *Science*, vol. 272, 21 June, 1996). Also, Frederick Seitz published an editorial-page piece in *The Wall Street Journal* (12 June, 1996) titled "A Major Deception on 'Global Warming'", which represented a frontal attack on the IPCC. Seitz maintained that "key changes" were made to the assessment report after the scientists had met and accepted "what they thought was the final peer-reviewed version". According to Seitz, more than 15 sections of the chapter were changed or deleted. Moreover, he maintained that these changes were not "merely cosmetic"; "nearly all worked to remove hints of the scepticism with which many scientists regard claims that human activities are having a major impact on climate in general and global warming in particular". His criticism was particularly directed against the convening lead author of this section, Benjamin D. Santer:

> I am in no position to know who made the major changes in Chapter 8; but the report's lead author, Benjamin D. Santer, must presumably take the major responsibility. ... Whatever the intent was of those who made these significant changes, their effect is to deceive policy makers and the public into believing that the scientific evidence shows human activities are causing global warming. (Cited from the reprint of the letter in Bulletin of the American Meteorological Society, 1996: 1963)

It should perhaps also be mentioned that Seitz had not taken part in the IPCC process, he had not participated at the meeting in question and he had not contacted IPCC lead authors or the WG or panel leadership to hear their version of the story before he launched his frontal attack in *The Wall Street Journal*.

The criticism was met by the convening lead author himself, as well as the WGI chair and the panel chair. Santer emphasised that the changes were made to improve the scientific clarity of the report and that the key phrase of the report was approved by all 100 participating governments (*Nature*, 13, vol. 381, June, 1996). Moreover, in a response to Seitz, Santer and 40 other scientists (all of whom are lead authors to the IPCC second assessment report) emphasise that the changes made to the report after the Madrid meeting were made in response to written comments by scientists, governments and non-governmental organisations, both from the review process and at the meeting. Above all, these scientists emphasise that the basic content of the chapter remains the same after the editing:

> The pre- and post-Madrid versions of the chapter are equally cautious in their statements. Uncertainties have not been suppressed. Roughly 20% of Chapter 8 is devoted to the discussion of uncertainties in estimates of natural climate variability and the expected 'signal' due to human activities.
>
> The deletions quoted by Seitz relate to the difficulties involved in attributing climate change to the specific cause of human activities, and to uncertainties in estimates of natural variability. These issues are dealt with at great length in the published chapter. The basic content of these particular sentences has not been deleted. (Letter from Ben Santer and 40 other IPCC scientists, printed in *The Wall Street Journal*, 25 June 1996, cited from reprint in Bulletin of the American Meteorological Society, 1996: 1963.)

In their response to Seitz's accusations, the WGI leadership and the panel chairman maintain that Seitz's criticism is based on a miscomprehension of the nature of the IPCC process. They maintain that the Madrid meeting and the subsequent revisions constituted the final part of the review process, and that the report was accepted by the WGI plenary on condition that these modifications were made. Hence, according to the leadership of the WG and the panel chair, the IPCC rules of procedure had been violated had these changes *not* been made (letter from Bolin, Houghton and Meira Filho printed, in part, in *The Wall Street Journal*, 25 June 1996, referred from reprint in Bulletin of the American Meteorological Society, 1996: 1965; see also *Nature*, vol. 381, 6 and 13 June 1996).

The inappropriateness of newspaper columns as arena for scientific debate is also indicated in the exchange over this issue which took place in *The Wall Street Journal*. The editor of the newspaper found it necessary to make quite substantial changes to the responses of both the IPCC lead authors and the WG and panel leadership, in the form of both deletions and additions. The *Journal*'s editing is clearly indicated in the reprint of the entire correspondence in the Bulletin of the American Meteorological Society (1996: 1961–66). For instance, with the exception of Ben Santer, all the names of the additional 40 lead authors who originally had signed the letter are deleted. The response from Bolin, Houghton and Meira Filho was even more heavily edited, with only two paragraphs of the original seven retained in the published version.

The situation that arose in the aftermath of the acceptance and publication of WGI's second assessment report was unfortunate. Moreover, the situation, both in the re-editing and in the debate that followed, was not handled very well by the WG and panel leadership. First, the re-editing itself should perhaps not have taken place at all, after the WGI plenary's acceptance of the assessment report. Even if, as the leadership maintains, this was part of a final incorporation of review comments and thus in accordance with the rules of procedure, the editing does leave the impression that the conclusions of the *assessment* are revised according to its *summary*. Given the amount of time spent in the WGI plenary on the notorious sentence of the SPM ("The balance of evidence suggests..."), the WG and panel plenary should have foreseen the problems such a re-editing would cause. Second, the re-editing does not at all seem worth the trouble: As pointed out in chapter 6, the corresponding sentence of the *draft* SPM is almost identical, and the re-wording of this sentence in the WGI plenary meeting would thus not have had any implications for the correspondence between the assessment as it was and the summary as it was re-worded in the plenary meeting.

The WG and panel leadership were, however, in a dilemma. The re-editing of the chapter was the final stage of the review process, with the incorporation of comments brought forward in the review process and at the WGI plenary meeting. Some of the re-editing was also done in response to comments from the same group that subsequently were its main critics (the oil-industry lobby). Thus, it may seem as if the WG and panel leadership would become entangled in a dispute with this group regardless of what they did.

The IPCC leadership may be criticised for handling this situation inappropriately. It is, however, also important to recognise that the debate concerned the question of whether the rise in temperature observed over the last decade may be attributed human activities. It does not concern the more general question of whether human activities can affect climate. The former question is much more controversial and disputed within the scientific community than the latter, and should perhaps, for that reason, not have been included in the summary of the assessment (although the "attribution" implied in the sentence, after all, is very cautious). The point remains, however, that even if this "attribution" actually is wrong, it does not have any significant implications for the main conclusion of the assessment: that human activities can affect global climate. This suggests, therefore, that the error of judgement by the WG and panel leadership that, in part, caused this situation, does not qualify as a (conscious or unconscious) deception of policymakers. The main conclusion of the assessment that human activities can affect global climate, and the scientific uncertainties associated with timing, magnitude and regional variations, are discussed throughout the assessment. Also, in the particular chapter in question, main uncertainties are thoroughly discussed.

8.3.3 In Sum

The IPCC's handling of the sign of the water-vapour feedback effect in the first assessment and their handling of the situation that arose in relation to the second assessment was unfortunate and inappropriate. The implications of these incidents for policymakers' comprehension of the state of knowledge, however, are not sufficiently significant for these incidents to be considered a misrepresentation of scientific uncertainty which may have led to policymakers' miscomprehension of the level of scientific uncertainty with which the knowledge base is associated. Overall, therefore, while the IPCC has faltered in its communication of scientific uncertainty on some occasions, there is no evidence indicating that policymakers have acted upon a misconception of the state of knowledge as a result of these mistakes by the IPCC.

It should perhaps also be noted that some of the criticism against the IPCC is based on a miscomprehension of the role of consensus in the assessment process. The consensus rule with which the reports are approved and accepted by the WG and full panel plenaries has been taken to indicate that IPCC scientists claim that climate change science is consensual – in the meaning of uncontroversial and conclusive. Lindzen, for instance, has commented on this "alleged" scientific consensus: "Why, one might wonder, is there such insistence on scientific unanimity on this issue. After all, unanimity in science is virtually non-existent on far less complex matters. Unanimity on an issue as uncertain as 'global warming' would be surprising and suspicious" (Lindzen, 1992: 7). Rather, as discussed in chapters 6 and 7, the consensus of the IPCC concerns the manner in which state-of-the-art climate change knowledge – *with* its inherent uncertainties and controversies – should be represented in IPCC reports and publications. Hence, the consensus of the IPCC is also a consensus regarding what constitutes major uncertainties and scientific controversies.

It should, however, also be emphasised that one source of this miscomprehension may be found in the IPCC reports themselves. Again, in the First Assessment, the self-congratulating (and unsigned) "pre-preface" is a particular case in point: "The information presented here is of the highest quality. It will inform the necessary scientific, political and economic debates and negotiations that can be expected in the immediate future. Appropriate strategies in response to the issue of climate change can now firmly be based on the scientific foundation that the Report provides" (IPCC, 1990).

8.4 Causal Relationship: Real?

The analysis suggests that the outcome in this case – in terms of the extent to which policymakers have accepted the scientific knowledge base provided by the IPCC – cannot be reproduced either by factors related to the political setting of the problem or by factors related to the state of knowledge in relevant fields of research without taking into account factors linked to the science–policy dialogue. We have described a political setting that clearly indicates that the problem can in no terms be characterised as politically benign. Rather, it is characterised by incompatible interests and positions and complex intra-national, intra-regional and international negotiations. As a field of research, moreover, the problem of a human-induced climate change is associated with significant scientific uncertainty. While the IPCC has faltered in some regards in their communication of this scientific uncertainty, there is no evidence of attempts by the IPCC to manipulate the conclusiveness of state-of-the-art knowledge which may have led policy makers to believe that the knowledge is more conclusive than it actually is. On this basis, therefore, it can be concluded that the validity of propositions A1 and A2 (a and b) – as alternative explanations to the outcome in this case – is low. Thus, the outcome in this case, and the level of effectiveness indicated in the policy response, is difficult to explain by means of the control variables of this study. This increases the likelihood that there is a causal relationship between institutional design and leadership behaviour in the science–policy interaction process on the one hand, and policy makers' acceptance of the knowledge base provided on the other. The relationship between science and politics and, indeed, the policymaking process whereby a decision on whether or not to act on an (accepted) problem diagnosis is made, is complex. This conclusion, therefore, should not be interpreted to indicate that factors related to the science–policy dialogue of the IPCC process are the only factors that can explain this outcome. On the contrary, it is highly unlikely that factors which may serve to enhance the effectiveness of the science–policy dialogue by themselves can explain the outcome of processes that are essentially political in nature (whether or not to act upon a scientific and a political problem diagnosis). While the empirical analysis suggests that there is a causal relationship between these factors and the extent to which policymakers have accepted the knowledge base provided by the IPCC, this does not exclude the possibility of equally important independent effects by other factors (such as, for instance, the public saliency of the issue and the role of NGOs in the policymaking process) and interaction effects between several factors. In particular, the explanatory power of interaction effects between (an effective communication of) scientific knowledge and political interests can be assumed to be significant (see, for instance, Litfin, 1994).

It is in the power of policymakers to act on the problem of a human-induced climate change. In general, policymakers may be assumed to be rational actors who act on the basis of their perceptions of the interests of their nations. We may thus assume that states – unless they are exposed to substantial external pressures – act in accordance with their interests in the problem area. What constitutes interests, however, and particularly how these interests are prioritised is, *inter alia*, a function of the terms in which problems are framed, defined and understood. Hence, the extent to which an effective communication of scientific knowledge contributes to a political redefinition of problems by increasing policy makers' awareness of how their realisation of political interests can be linked or combined with measures to resolve or abate particular problems will also determine the extent of the scientific impact on the policy area concerned. There is no reason to believe that the process from a scientific problem definition to a political acceptance of the problem and a political problem definition is linear and chronological in nature. Rather, several studies have pointed to the simultaneousness characterising this process (see, for instance, Litfin, 1994; Hajer, 1995). In this perspective, it is probable that the extent to which policy makers accept the knowledge base provided by the scientific community as factually valid also is determined by the extent to which the scientific problem definition helps them to frame and define the problem so that it is compatible with their perception of interests. This does, however, imply that the science–policy dialogue is effective.

The extent to which the scientific problem definition contributes to a redefinition of interests among powerful interest groups at the national level may be equally important. For instance, when the chemical industry realised that the problem of ozone depletion would not go away by itself and that they, sooner or later, would have to face it, they also realised that it might be profitable to react sooner rather than later (see, for instance, Maxwell and Weiner, 1993). While there are a set of factors that may have influenced the chemical industry's redefinition of interests in relation to the problem of ozone depletion, their participation in the scientific assessment process is likely to have played a significant role.

So far, the major interest group associated with the climate change problem, the energy lobby particularly in the United States, has represented one of the most significant "obstacles" to more offensive climate change policies. The last couple of years, however, have witnessed an interesting developing trend which may indicate a slow progress towards a redefinition of interests also within this group. In 1996, during the turbulence created by the GCC's accusations against the IPCC for "scientific cleansing", two major companies, British Petroleum (BP) America and the Arizona Service Company, withdrew from the coalition. In April 1998, The Royal Dutch/Shell Group did the same. In May 1997, BP CEO John Browne stated that "there is now an effective consensus among the world's leading scientists and serious and well-informed people outside the scientific community that there is a discernible human influence on global climate". He further stated that "we need to go beyond analysis to seek solutions and take action," calling it "a moment for change and rethinking of corporate responsibility" on the environment (*GECR*, 13 June 1997). On this occasion, John Browne also announced BP's plans to actively develop alternative fuels, and particularly the development of competitive solar power technology. Thus, BP's aspiring redefinition of its interests may also reflect a redefinition of its function – from defining itself as an oil company towards a redefinition of itself as an energy company.

While it is too early to judge the significance of this development, it certainly gives some indication of what constitutes an effective communication of scientific knowledge and the interaction effects that may characterise the relationship between scientific knowledge and political as well as commercial interests. Thus, the conclusion that there is a causal relationship between the way the science–policy dialogue is organised, the performance of individual leadership in the science–policy interaction process, and policymakers' acceptance of the knowledge base in this case does not exclude the possibility of such interaction effects. On the contrary, it may serve as a point of departure for delving into these complex interrelationships to further improve the knowledge about available tools to facilitate processes of international environmental regime formation.

Chapter 9

Structure and Agent in the Scientific Diplomacy of Climate Change

The proposition explored in this analysis – that institutional design may be utilised as a tool for improving the effectiveness of the science–policy dialogue – is fascinating both in its apparent simplicity and in the vast scope of opportunities it seems to imply. It is a fascinating thought that we may design institutions in accordance with knowledge about how institutional arrangements work and thereby improve their performance. As expressed by E. B. Haas, "it is tempting to design perfect institutions" (1990: 177).

As shown in this and other studies, however, the mechanisms that link institutional arrangements and institutional performance are characterised by complexity. For instance, social institutions are not always *designable.* Institutional arrangements may be consciously designed under certain conditions and within certain limits. Also, while the potential of institutional arrangements as tools for improving institutional performance lies in their impact on actor behaviour, actor behaviour is also an important determinant to how institutional arrangements work. Thus, the relationship between the structure of social institutions and the agents operating within these structures is, above all, characterised by contingency.

The contingent relationship between institutional arrangements and actor behaviour is evident also in the case of the IPCC. In the theoretical approach to this analysis, it was suggested that the institutional arrangements of processes of science–policy interaction could be assumed to enhance the effectiveness of the process to the extent that they were capable of serving four main functions: (i) maintaining the scientific autonomy and integrity of scientific bodies; (ii) ensuring a certain level of science–policy involvement and interaction in the process; (iii) ensuring the geopolitical representativeness of the bodies; and (iv) providing mechanisms for conflict resolution. The empirical analysis revealed that the institutional apparatus of the IPCC has served these functions reasonably well, and that this indeed has served to enhance the effectiveness of the process.

In particular, it was found that the capacity of the institutional apparatus to combine scientific autonomy and science–policy integration is crucial for the success of the endeavour. In the IPCC, the combination of these seemingly incompatible functions proved possible because of the differentiation of institutional arrangements according to function in the assessment process. While scientific autonomy seems primarily to facilitate the development of the knowledge base, integration is a crucial element in the transformation of the knowledge base into decision-making premises. As has been demonstrated, these functions are served by different decision-making levels within the IPCC system: the provision of the knowledge base takes place within the scientific core, while its transformation into decision premises primarily takes place in the WG and panel plenaries. This differentiation of functions also required a differentiation of institutional arrangements. The scientific core is characterised by a relatively high level of scientific autonomy and a low level of science–policy integration and interaction, whereas the WG and panel plenaries are characterised by a low level of scientific autonomy and a relatively high level of science–policy integration. Thus, through this differentiation of roles and functions in the assessment process, a combination of scientific autonomy and science–policy integration is achieved within each WG of the IPCC.

This institutional set-up implies that the knowledge base is subjected to intense scientific and political scrutiny throughout the process. In the development of the knowledge base, scientific scrutiny is ensured through extensive review procedures. In its transformation, the knowledge base (the assessment summaries) is submitted to adversarial scrutiny by parties representing conflicting interests in the policy area. Thus, the IPCC has succeeded in developing a knowledge base constructed by scientists respected and recognised for their competence and integrity as well *as* in submitting this knowledge base to the adversarial scrutiny by parties representing conflicting interests in the policy area. The combination of these features of the IPCC process seems to have contributed substantially to enhancing the acceptability of the knowledge base provided to all "stakeholders" – scientists as well as policymakers. Interestingly, both of these mechanisms seem to have enhanced policymakers' sense that the knowledge base provided is impartial. Thus, in a paradoxical sense, the participation of policymakers in the assessment process seems to have contributed to their acceptance of the knowledge base as impartial and disinterested, and hence, as a value-free and scientifically authoritative problem diagnosis.

The effect of the institutional arrangements of the IPCC process is also, however, found to be strongly correlated with the provision of a set of crucial leadership functions in the process. This correlation is particularly well illustrated with regard to the provision of mechanisms for conflict resolution in the process. The institutional arrangements of the IPCC process do not contribute directly to the provision of mechanisms for conflict resolution. While there are institutional arrangements that may be utilised as tools for the resolution of conflicts, they are subtle and they are also designed for other purposes in the assessment process. At the same time, the concern for critical review and adversarial scrutiny which penetrates the assessment process at all decision-making levels of the IPCC is a conflict generating procedure. This mode of operation has thus served to enhance the demand for instruments to handle and resolve disputes. This function was largely served by agents operating within the system. As has been shown, the WG plenaries walk a thin line between effective progress and complete deadlock. The eventual outcome seems to a very large extent to have depended upon the substantive and personal capabilities of the WG and panel chairs to provide firm leadership. The proceedings of the WG plenaries are characterised by intense science–policy and policy-internal discussions, with a significant element of strategic reasoning, often approaching straightforward negotiations. Operating under a more or less constant threat of complete deadlock, the contribution by the WG and panel chairs to the development of a knowledge base which is acceptable to all stakeholders is a piece of scientific diplomacy well done. The outcome could very well have been completely different had this crucial leadership function not been provided.

The contingent relationship between institutional arrangements and actor behaviour is thus very well illustrated in this case. The analysis provided here is not only evidence of the independent effect of two independent variables on the outcome of the process, it strongly suggests that the impact of each of these variables is contingent upon the value of the other. In particular, a completely different (contrary) effect of the institutional arrangements of the IPCC process can be envisaged had these crucial leadership functions not been provided. This suggests that an important determinant of the effectiveness of processes of science–policy interaction is the complementarity and contingency characterising the relationship between these two factors. How institutional arrangements work depends to a large extent upon how agents operating within the system make them work.

An interesting aspect of this analysis, related to the contingency characterising the relationship between institutional arrangements and actor behaviour, is the inverse relationship that it suggests between the effectiveness of institutional arrangements and the robustness of this effect. The empirical analysis suggested that scientific autonomy and science–policy integration might not be as hard to combine as initially assumed. Having analysed in more detail the role of leadership in this process, however, it may seem that the incompatible relationship between autonomy and integration is replicated in the inverse relationship between the effectiveness of institutional arrangements and the robustness of this effect. This suggests that the more effective institutional arrangements may be, the less robust is this effect. This implies that "effective" institutional arrangements are also fragile, context-dependent and contingent upon less controllable factors such as the provision of leadership functions. As we have seen, a successful provision of leadership depends not only on the formal and substantive capabilities of individual actors, but also in large part on these individual actors' personal capabilities or interpersonal skills. The provision of leadership, therefore, can only to a very little extent be ensured, for instance, through institutional arrangements such as recruitment procedures. This is, above all, indicated by the role of lead authors in the development of the assessments. Lead authors operate in a role which perhaps most clearly is defined in terms of formal skills – their scholarly merits. Still, the situation that arose after the shift of lead author in WGI in the middle of the preparation of the First Assessment report indicates that scientific brilliance is not enough to perform a successful lead authorship. A crucial part of this role seems to be the personal capacity of the lead author to handle dispute and conflict in the assessment process and to induce co-operation among the participants to the process. Even in the case of lead authors, therefore, the role requires personal capabilities that cannot be ensured through formal recruitment procedures. This is even more true with regard to WG chairs, whose role to a lesser extent is based on clearly defined formal and substantive skills, and whose function in the assessment process is of key importance to the effect of institutional arrangements.

One main conclusion that can be drawn from this analysis, is therefore, that the effectiveness of processes of science–policy interaction is not exclusively determined by, and hence cannot be manipulated solely by design. The effect of institutional arrangements is, above all, dependent upon the provision of a set of leadership functions which cannot be ensured through formal procedures and institutional arrangements. It should be emphasised, however, that while the successful provision of leadership functions cannot be ensured through institutional design, a great deal can be done to encourage and provide incentives for this mode of behaviour. The provision of a mandate for leadership is but one such institutional device. Our inability to develop perfect institutions by design does not imply, moreover, that the way in which processes of this kind are organised is unimportant for the extent to which they succeed in developing a scientific and political problem diagnosis. Rather, it is an indicator of the complexity characterising the relationship between institutional design and institutional performance in science–policy interaction and the amount of personal skill that is also involved in a successful endeavour.

In this analysis, a two-level conceptualisation of effectiveness has been employed. Both the institutional design – the structure of the IPCC – and the provision of leadership – the behaviour of agents in the process – are factors that have been found to have contributed substantially to the development of a problem diagnosis that is accepted and agreed upon by both scientists and policymakers. The successful performance of the IPCC process has, however, not been sufficient to induce policymakers to act upon this consensual problem diagnosis.[1] Thus the differentiated conceptualisation of effectiveness has enabled the identification of factors serving to enhance the effectiveness of the science–policy dialogue, even in a case which is not optimally effective. In this regard, the analytical distinction between levels of effectiveness has been useful also in the sense that it has enabled us to envisage the limits within which the effectiveness of the science–policy dialogue can make a difference for the extent to which policymakers choose to respond to available scientific knowledge in terms of actions. This analytical distinction thus also gives an indication of the scope for utilising the institutional design of the science–policy dialogue as an instrument to enhance the effectiveness of the process. While the institutional arrangements of the science–policy dialogue may have a direct impact on the extent to which a consensual problem diagnosis is developed, they do not have a direct impact on the extent to which policymakers act upon this knowledge. This implies, therefore, that institutional design represents a tool for enhancing the effectiveness of the process primarily in terms of the extent to which policymakers accept the knowledge base provided, preferably in the form of a consensual problem diagnosis. Beyond the fact that policymakers' acceptance of the factual validity of the knowledge base is a necessary first requirement for their (subsequent) adoption of the knowledge base as a basis for actions, the effectiveness of the science–policy dialogue seems to have little impact on whether and when policymakers choose to act upon available knowledge. Even in processes where the science–policy dialogue has functioned very well, therefore, policymakers may choose not to respond in terms of actions for reasons beyond which the

[1] The adoption of the Kyoto Protocol is not considered to represent a behavioural change which indicates that policymakers have acted upon the policy implications of the knowledge brought forward by the IPCC.

effectiveness of the science–policy dialogue can make a difference. This confirms the rather obvious assumption that the provision of a problem diagnosis is only but one determinant of international environmental regime formation. It need not, even, be the most important.

The extent to which policymakers have accepted the factual validity of the knowledge base provided by the IPCC, as indicated by the Kyoto agreement, is difficult to explain by the control variables analysed in this study. As a problem of incongruity, the political malignancy of the climate issue is significant. This indicates that pure political opportunism is unlikely to have produced this outcome. Also, the state of knowledge in this area is characterised by a significant level of scientific uncertainty, which, as far as can be judged (apparently) has been discussed and communicated to policymakers in the IPCC assessments. This indicates that the incentive to act which is inherent in conclusive knowledge also is unlikely to have produced this outcome. This does not imply that other factors, not investigated in this analysis, have not contributed to the outcome. Factors such as the public saliency of the issue and the role of environmental NGOs may very well have been at work in this case and contributed to this outcome. The inability of the control variables of this study to produce this outcome is, however, considered to be a strong indication that the factors investigated here – the institutional design of the process and the provision of leadership functions – constitute important explanatory factors.

While the political setting of the problem and the state of knowledge in relevant research fields are unlikely candidates to explain the extent to which policymakers have accepted the knowledge base provided by the IPCC, they are most likely candidates for explaining the (low) extent to which policymakers have acted upon the policy implications of this acceptance. As we have seen, the climate change problem is politically complex and difficult not only because it is a collective-action problem and, hence, a problem of incongruity. The development of political solutions to the problem of a human-induced climate change is difficult also because of the costs involved and the difficult burden-sharing issues, particularly between the North and the South, that are generated. The climate issue is also riddled with scientific uncertainty – an uncertainty which is likely to last for decades to come. Policymakers thus have to confront and deal with this uncertainty in their design of political solutions. Moreover, scientific uncertainty interacts with the political malignancy of the problem, particularly in the sense that cost–benefit analyses that reflect the true costs and benefits of both action and inaction are almost impossible to make. In this regard, it could be argued that the level of policy action resulting from the negotiation process is an achievement in the face of these complicating and complex conditions. It also suggests that other factors have been at work in this process.

As a field of research within political science, the mechanisms at work in processes of science–policy interaction in international environmental regime formation and the factors that make one endeavour work differently from another, is in a sense characterised by theoretical "poverty". The main, and an important, contribution to enhanced insight into processes of this kind is the epistemic communities approach (P. M. Haas, 1989a,b; 1992a,b; 1993a,b). The epistemic communities approach, however, takes as its point of departure a basis of consensual knowledge, and explores the role of "epistemic communities" as one main vehicle for the communication and diffusion of this knowledge to policymakers. Thus, this approach offers very little in terms of exploring the mechanisms at work in the development of consensus between and among scientists and policymakers, and can to a very little extent offer alternative explanations in cases where epistemic communities have not been at work. Also, the epistemic communities approach does not explore or investigate the transformation of knowledge into decision premises and the mechanisms at work under conditions of scientific uncertainty. A significant contribution to the exploration of these aspects has been provided by Litfin and her investigation of "knowledge brokers" (1994). In her analysis of the science–policy discourse associated with the ozone regime, she has pointed to the power of language in the formulation and interpretation of scientific facts under conditions of scientific uncertainty, and the potential role of intermediaries – knowledge brokers – in the transformation of pure knowledge (as it appears, for instance, in scientific journals) into points of information that are comprehensible and, not least, applicable to policymakers. While Litifin's study generates significant insight into the functions served by knowledge brokers in processes of science–policy interaction and the mechanisms at work that provide knowledge brokers with their influence, her approach can to a very little extent offer explanations in cases where knowledge brokers have not been at work. Also, both of these approaches have little to offer in terms of what practitioners themselves can do to affect the course of the process and how they may increase the likelihood of success.

This analysis does not, and has not sought to, invalidate any of these approaches to processes of science–policy interaction. Thus, it does not have theoretical implications in the direction of falsification. It has, however, approached the question of effectiveness in the dialogue between scientists and policymakers in broader and more general terms. It has also sought to develop a perspective that can provide insight into the mechanisms at work in processes of this kind, which may be useful to identify instruments to facilitate the process that may be available to practitioners themselves. Thus, the analysis seeks to contribute to the development of a more fine-tuned and detailed understanding of the mechanisms at work in processes of science–policy interaction. For instance, by focusing on the multitude of potential pitfalls of processes of this kind, and the ways in which institutional design may be utilised as an instrument for coping with these obstacles, this analysis also offers a broader set of possible explanations of failure.

While the epistemic communities and the knowledge-brokering approaches both investigate processes of science–policy interaction in terms of the behaviour of specific individuals or groups of individuals, this analysis has focused on the structural framework within which agents operate. The analysis has, therefore, generated insight also into some of the conditions under which epistemic communities, knowledge brokers and/or, individual actors who take on various leadership roles are likely to succeed in their endeavour. In this regard, the approach developed here encompasses the epistemic communities and the knowledge-brokering approaches. The conclusion that the relationship between structure and agent, above all, is characterised by contingency, and that the behaviour of individual agents is a crucial determinant to the success of the process, does, for instance, at least in an indirect sense, support the findings of these approaches. Thus, rather than invalidating these approaches, this analysis has supplemented them. It can be hoped, however, that the findings of this analysis also contribute to sensitising individual agents to the tasks for which their contribution is required for the process to succeed and contribute to an increased understanding of the institutional devices that may be utilised to induce agents to take on these crucial roles.

Scientific knowledge may play several roles in policymaking; it may be a guide to policy choice, but it may also represent an instrument of deception. Moreover, the features of scientific knowledge that make it qualified as a guide to action are also the source of its potential as an instrument of deception. This complex relationship seems to suggest that a strict separation of science and policy is required to avoid the most paralysing pitfalls and most dysfunctional dynamics inherent in science–policy interaction. Scientific knowledge in its pure form is not, however, readily available for policymakers to apply. Indeed, one main objective of the process is to define and diagnose problems not only in their scientific terms, but also within their political context. In the development of a scientific and political problem diagnosis, the competence of both scientists and policymakers is needed. It is this duality in the requirements for a successful science–policy dialogue that is the source of the main challenge facing both organisers of and participants in processes of this kind. This analysis suggests that their ability to handle the dilemmas involved in the balancing of science–policy separation and integration is one significant determinant of the success of their endeavour.

References

Agrawala, S. (1998a). 'Context and Early Origins of the Intergovernmental Panel on Climate Change', forthcoming in *Climatic Change*.

Agrawala, S. (1998b). 'Structural and Process History of the Intergovernmental Panel on Climate Change', forthcoming in *Climatic Change*.

Andresen, S. and W. Østreng (eds.) (1989), *International Resource Management: The Role of Science and Politics*. London and New York: Belhaven Press.

Andresen, S., T. Skodvin, A. Underdal and J. Wettestad (1994), *"Scientific" Management of the Environment? Science, Politics and Institutional Design*. CICERO, Working Paper 1994:4.

Andresen, S., T. Skodvin, A. Underdal and J. Wettestad (in press), *Science in International Environmental Regimes: Between Integrity and Involvement*. Forthcoming at Manchester University Press.

Bacharach, S. B., and E. L. Lawler (1988), *Bargaining: Power, Tactics, and Outcomes*. San Francisco: Jossey-Bass Publishers.

Benedick, R. E. (1991), *Ozone Diplomacy: New Directions in Safeguarding the Planet*. Cambridge, Massachusetts: Harvard University Press.

Beuermann, C. (1997), China and Climate Change. In G. Fermann (ed.), *International Politics of Climate Change: Key Issues and Critical Actors*. Oslo: Scandinavian University Press, pp. 213-235.

Boehmer-Christiansen, S. (1993), Scientific Uncertainty and Power Politics: The Framework Convention on Climate Change and the Role of Scientific Advice. Paper prepared for the conference: *"Geopolitics of the Environment and the New World Order"*, Chantilly, January 1993.

Boehmer-Christiansen, S. (1994), Global Climate Protection Policy: The Limits of Scientific Advice, Part 1 and 2, *Global Environmental Change*, 4, no.2: pp. 140-159, and no.3: pp. 185-200.

Boehmer-Christiansen, S. (1995), Britain and the International Panel on Climate Change: The Impacts of Scientific Advice on Global Warming. Part I: Integrated Policy Analysis and the Global Dimension, *Environmental Politics*, vol. 4, no.1, pp. 1-18.

Bolin, B. (1998). 'The Kyoto Negotiations on Climate Change: A Science Perspective'. *Science*, 279: 16 January.

Bolin, B., J. Houghton, and L. G. M. Filho (1996). 'Open Letter to Ben Santer'. *Bulletin of the American Meteorological Society*, 77: 1965-1966.

Borione, D. and Ripert, J. (1994). 'Exercising Common but Differentiated Responsibility', in I. M. Mintzer and A. J. Leonard (eds.), *Negotiating Climate Change: The Inside Story of the Rio Convention.* Cambridge: Cambridge University Press, pp. 77-97.

Brenton, T. (1994). *The Greening of Machiavelli: The Evolution of International Environmental Politics*. London: Earthscan Publications.

Buzan, B. (1981), 'Negotiating by Consensus: Developments in Technique at the United Nations Conference on the Law of the Sea' *American Journal of International Law*, vol. 72, no. 2; pp. 324-348.

Carter, T. R., M. L. Parry, H. Harasawa and S. Nishioka (1994), *IPCC Technical Guidelines for Assessing Climate Change Impacts and Adaptations*. London: Department of Geography, University College London.

Cess, R. D. (1991), Positive about Water Feedback. *Nature*, vol. 349, 7 February, pp. 462-463.

Cole, S. (1992), *Making Science: Between Nature and Society*, Cambridge Mass.: Harvard University Press.

Collingridge, D. and C. Reeve, (1986), *Science Speaks to Power: The Role of Experts in Policy Making*, London: Frances Pinter Publishers.

Cubasch, U. and R. D. Cess (1990), Processes and Modelling. Chapter 3 in J. T. Houghton et al., *Climate Change: The IPCC Scientific Assessment. Report Prepared for IPCC by Working Group I.* Cambridge: Cambridge University Press, pp.69-93.

Dasgupta, C. (1994), The Climate Change Negotiations. In I. M. Mintzer and J. A. Leonard (eds.), *Negotiating Climate Change: The Inside Story of the Rio Convention*. Cambridge: Cambridge University Press, pp. 129-149.

Dickinson, R. E., V. Meleshko, D. Randall, E. Sarachik, P. Silva-Dias and A. Slingo (1995), Climate Processes. In J. T.Houghton et al., *Climate Change 1995: The Science of Climate Change*. Cambridge: Cambridge University Press, pp. 193-229.

Effinger, M. and M. Zürn (1990), Explaining Conflict Management in East–West Relations: A Quantitative Test of Problem-Structural Typologies. In V. Rittberger (ed.), *International Regimes in East–West Politics*. London and New York: Pinter Publishers, pp. 64-90.

Egeberg, M. (1984), *Organisasjonsutforming i offentlig virksomhet*. Oslo: Aschehaug/Tanum-Norli.

Elster, J. (1985), 'Rationality, Morality and Collective Action' *Ethics*, vol. 96, no. 1; pp. 136-156.

Evans, Peter B., Harold K. Jacobsen og Robert D. Putnam (red.) (1993), *Double-Edged Diplomacy: International Bargaining and Domestic Politics*. Berkeley and Los Angeles: University of California Press.

Fermann, G. (1997), Political Leadership and Climate Change: The Prospects of Germany, Japan and the United States. In G. Fermann (ed.), *International Politics of Climate Change: Key Issues and Critical Actors*. Oslo: Scandinavian University Press, pp. 341-407.

Fisher, R. and W. Ury (1981/87), *Getting to Yes: Negotiating Agreement Without Giving In*. London: Arrow Books Limited.

Fuglestvedt, J. S., T. Berntsen, O. Godal and T. Skodvin (subm.), Climate Implications of GWP-based Reductions in Greenhouse Gas Emissions. Submitted to Geophysical Research Letters, June, 1999.

Gates, W. L., J. F. B. Mitchell, G. J. Boer, U. Cubasch and V. P. Meleshko, (1992), Climate Modelling, Climate Prediction and Model Validation. In J. T. Houghton et al., *Climate Change 1992: The Supplement Report to the IPCC Scientific Assessment*. Cambridge: Cambridge University Press, pp. 97-135.

Gelbspan, R. (1997), *The Heat is On: The High Stakes Battle over Earth's Threatened Climate*. Massachusetts, California and New York: Addison-Wesley Publishing Company.

Gibbons, M.; C. Limoges; H. Nowotny; S. Schwartzman; P. Scott; and M. Trow, (1994), *The New Production of Knowledge: The Dynamics of Science and Research in Contemporary Societies*. London: Sage Publications.

Gilbert, N. (1976), The Transformation of Research Findings into Scientific Knowledge, *Social Studies of Science*, vol. 6, no. 3 & 4, pp. 281-307.

Global Environmental Change Report (GECR), 11 August, 1995; 'IPCC Social Cost Assessment Sparks Controversy'.

Global Environmental Change Report (GECR), 8 September, 1995: 'IPCC Debates Economic Models'.

Global Environmental Change Report (GECR), 13 June, 1997: 'Shell Withdraws from Global Climate Coalition'.

Grubb, M. (1995), European Climate Change Policy in a Global Context. In H. O. Bergesen and G. Parmann (eds.), *Green Globe Yearbook 1995*. Oxford: Oxford University Press, pp. 41-51.

Gupta, J. (1997), *The Climate Change Convention and Developing Countries: From Conflict to Consensus?* Dordrecht: Kluwer Academic Publishers.

Haas, E. B. (1975), Is There a Hole in the Whole? Knowledge, Technology, Interdependence, and the Construction of International Regimes. *International Organization*, vol. 29, no. 3, pp. 827-876.

Haas, E. B., M. P. Williams and D. Babai (1977), *Scientists and World Order: The Uses of Technical Knowledge in International Organizations*, Berkeley: University of California Press.

Haas, E. B. (1980), Why Collaborate? Issue-Linkage and International Regimes, *World Politics*, vol. 32, no. 3, pp. 353-405.

Haas, E. B. (1990), *When Knowledge is Power: Three Models of Change in International Organizations*. Berkeley, Los Angeles and London: University of California Press.

Haas, P. M. (1989a), Do Regimes Matter? Epistemic Communities and Mediterranean Pollution Control, *International Organization*, 43, no.3, Summer 1989.

Haas, P. M. (1989b), *Saving the Mediterranean: The Politics of International Environmental Cooperation*, New York: Columbia University Press.

Haas, P. M. (1992a), Introduction: Epistemic Communities and International Policy Coordination, *International Organization*, vol. 46, no.1, pp. 1-37.

Haas, P. M. (1992b), Banning Chlorofluorocarbons: Epistemic Community Efforts to Protect Stratospheric Ozone, *International Organization*, vol. 46, no.1, pp. 187-225.

Haas, P. M. (1993a), Stratospheric Ozone: Regime Formation in Stages, in O. R. Young and G. Osherenko (eds.), *Polar Politics: Creating International Environmental Regimes*, Ithaca and London: Cornell University Press, pp. 152-186

Haas, P. M. (1993b), Epistemic Communities and the Dynamics of International Environmental Co-Operation, in V. Rittberger and P. Mayer (eds.), *Regime Theory and International Relations*, Oxford: Claredon Press, pp. 168-202.

Hajer, M. A. (1995), *The Politics of Environmental Discourse: Ecological Modernization and the Policy Process*. Oxford: Clarendon Press.

Houghton, J. T., G. J. Jenkins and J. J. Ephraums (eds.) (1990), *Climate Change: The IPCC Scientific Assessment. Report Prepared for IPCC by Working Group I.* Cambridge: Cambridge University Press.

Houghton, J. T. et al., (eds.) (1992), *Climate Change 1992: The Supplement Report to the IPCC Scientific Assessment.* Cambridge: Cambridge University Press.

Houghton, J. T. et al., (eds.) (1994), *Climate Change 1994: Radiative Forcing of Climate Change.* Cambridge: Cambrdige University Press.

Houghton, J. T. et al. (eds.) (1995), *Climate Change 1995: The Science of Climate Change.* Cambridge: Cambridge University Press.

Intergovernmental Panel on Climate Change (1990), *Climate Change: The IPCC Response Strategies.* World Meteorological Organization/ United Nations Environment Program. Washington D. C.: Island Press.

Intergovernmental Panel on Climate Change, Panel Plenary session reports from 1st to 13th session, 1988-1997.

Jasanoff, S. (1990), *The Fifth Branch: Science Advisers as Policymakers.* Cambridge, Mass.: Harvard University Press.

Keohane, R. O. (ed.), (1986), *Neorealism and its Critics.* New York: Columbia University Press.

Keohane, R. O. (1989), *International Institutions and State Power: Essays in International Relations Theory.* Boulder, San Francisco: Westview Press.

Keohane R. O., and J. S. Nye Jr. (1977), *Power and Interdependence: World Politics in Transition.* Boston, Toronto: Little, Brown and Company.

Keohane, R. O., and J. S. Nye Jr. (1987), Power and Interdependence Revisited. *International Organization*, vol. 41, no. 4; pp. 725-753.

Krasner, S. D. (ed.), (1983), *International Regimes.* Ithaca: Cornell University Press.

Krasner, S. D. (1988), Sovereignty: An Institutional Perspective. *Comparative Political Studies*, vol. 21, no. 1, pp. 66-94.

Kuhn, T. S. (1962/1970), *The Structure of Scientific Revolutions*, 2nd Edition, Chicago: The University of Chicago Press.

Kuhn, T. S. (1970a), Logic of Discovery or Psychology of Research? In I. Lakatos and A. Musgrave (eds.), *Criticism and the Growth of Knowledge.* Cambridge: Cambridge University Press, pp. 1-25.

Kuhn, T. S. (1970b), Reflections on my Critics. In I. Lakatos and A. Musgrave (eds.), *Criticism and the Growth of Knowledge*. Cambridge: Cambridge University Press, pp. 231-279.

Kuhn, T. S. (1977a), Second Thoughts on Paradigms. In F. Suppe (ed.), *The Structure of Scientific Theories*. Urbana, Chicago, London: University of Illinois Press, pp. 459-517.

Kuhn, T. S. (1977b) *The Essential Tension. Selected Studies in Scientific Tradition and Change*. Chicago and London: University of Chicago Press.

Kuhn, T. S. (1991), *The Trouble with the Historical Philosophy of Science*. Robert and Maureen Rothschild Distinguished Lecture, 19 November, 1991, Harvard University.

Lakatos, I. (1970), Falsification and the Methodology of Scientific Research Programmes. In I. Lakatos and A. Musgrave (eds.), *Criticism and the Growth of Knowledge*. Cambridge: Cambridge University Press, pp. 91-197.

Lakatos, I. and A. Musgrave (eds.) (1970), *Criticism and the Growth of Knowledge*. Cambridge: Cambridge University Press.

Latour, B. and S. Woolgar, (1979/1986), *Laboratory Life: The Construction of Scientific Facts*. New Jersey: Princeton University Press.

Lewin, L. (1988), *Det gemensamma bästa. Om egeninteresset och allmäninteresset i västerländsk politik.*Borås: Carlsson Bokförlag.

Lindzen, R. (1990a), Some Coolness Concerning Global Warming, *Bulletin of the American Meteorological Society*, vol. 71, No. 3, pp. 288-299.

Lindzen R. (1990b), Response from Comments by A. K. Betts. *Bulletin of the American Meteorological Society*, vol. 71, no. 10, October, pp. 1465-1467.

Lindzen, R. (1992), 'Global Warming: The Origin and Nature of Alleged Scientific Consensus', presentation at the *OPEC Seminar on the Environment*, Vienna, 13 - 15 April, 1992.

Lindzen, R. (1993), Letters to *Australian Meteorological and Oceanographic Society Bulletin*, April, pp. 22-24.

List, M. and V. Rittberger (1992), Regime Theory and International Environmental Management, in A. Hurrel and B. Kingsbury (eds.) *The International Politics of the Environment: Actors, Interests, and Institutions*. Oxford: Claredon Press, pp. 85-110.

Litfin, K., (1994), *Ozone Discourses: Science and Politics in Global Environmental Cooperation*. New York: Columbia University Press.

Litfin, K. (1995), Framing Science: Precautionary Discourse and the Ozone Treaties, *Millennium*, vol. 24, no.2, pp. 251-279.

Lunde, L. (1991). *Science or Politics in the Global Greenhouse? A Study of the Development towards Scientific Consensus on Climate Change* Lysaker: The Fridtjof Nansen Institute, EED Report 1991/8.

Mahlman, J. D. (1997), Uncertainties in Projections of Human-Caused Climate Warming. *Science*, vol. 278, 21 November, pp. 1416-1417.

Malnes, R. (1995), 'Leader' and 'Entrepreneur' in International Negotiations: A Conceptual Analysis. *European Journal of International Relations*, vol. 1, no. 1, pp 87-112.

March, J. G., and J. P. Olsen (1989), *Rediscovering Institutions: The Organizational Basis of Politics*. New York: The Free Press.

Masterman, M. (1970), The Nature of a Paradigm. In I. Lakatos and A. Musgrave (eds.), *Criticism and the Growth of Knowledge*. Cambridge: Cambridge University Press, pp. 59-91.

Maxwell, J. H. and S. L. Weiner (1993). 'Green Consciousness or Dollar Diplomacy? The British Response to the Threat of Ozone Depletion'. *International Environmental Affairs*, 5: 19-41.

McBean, G. and J. McCarthy (1990), 'Narrowing the Uncertainties'. Chapter 11 in J. T. Houghton, G. J. Jenkins and J. J. Ephraums (eds.), *Climate Change: The IPCC Scientific Assessment. Report Prepared for IPCC by Working Group I*. Cambridge: Cambridge University Press.

McKensie, D. (1989), How to use Science and Influence People, *New Scientist*, 29April, 1989.

Midgaard, K. (1976), Co-operative Negotiations and Bargaining: Some Notes on Power and Powerlessness, in Brian Barry (ed.), *Power and Political Theory: Some European Perspectives*, London: John Wiley.

Miles, E. (1977), 'The Structure and Effect of the Decision Process in the Seabed Committee and the Third United Nations Conference on the Law of the Sea' *International Organization*, 31; pp. 159-234.

Miles, E. (1989), Scientific and Technological Knowledge and International Cooperation in Resource Management. In S. Andresen and W. Østreng (eds.), *International Resource Management: The Role of Science and Politics*. London and New York: Belhaven Press. pp. 46-88.

Miles, E. et al., (subm.), *Explaining Regime Effectiveness: Confronting Theory with Evidence*.

Mitchell, J. F. B., S. Manabe, T. Tokioka and V. Meleshko (1990), Equilibrium Climate Change. Chapter 5 in Houghton et al., (eds.), *Climate Change: The IPCC Scientific Assessment. Report Prepared for IPCC by Working Group I*. Cambridge: Cambrdige Universtiy Press, pp. 131-173.

Mitchell, R., T. Bernauer (1998), Empirical Research on International Environmental Policy: Designing Qualitative Case Studies. *Journal of Environement and Development*, vol. 7, no.1, pp. 4-31.

Morrow, J. D. (1994), 'Modeling the Forms of International Cooperation: Distribution versus Information'. *International Organization*, vol. 48, no. 3, pp. 387-423.

Mulkay, M. (1976), The Mediating Role of the Scientific Elite, *Social Studies of Science*, vol.6, numbers 3 and 4: pp. 445-471.

Mulkay, M. (1978), Consensus in Science, *Sociology of Science*, vol. 17, no. 1, pp. 107-122.

Nature, vol. 371, 6 October, 1994: 'Discord over IPCC Meeting Reopens Climate Dispute', by D. Dickinson.

Nature, vol. 378, 9 November, 1995, p. 119; 'UN Climate Change Report Turns up the Heat', by E. Masood and A. Ochert.

Nature vol. 378, 30 November, 1995, p. 429; 'Temperature Rises in Dispute over Costing Climate Change', by E. Masood.

Nature vol. 378, 30 November, 1995, 433; 'Economics of Climate Change', by A. Meyer.

Nature vol. 379, 11 January, 1996, p. 108: 'Purpose and Function of IPCC', by M. Grubb.

Nature vol. 381, 6 June, 1996, p. 455: 'Head of Climate Group Rejects Claims of Political Influence', by E. Masood.

Nature, vol. 381, 13 June, 1996: 'Climate Report 'Subject to Scientific Cleansing'', by E. Masood.

Nature, vol. 390, 18/25 December 1997: 'Battling for Science Takes its Toll on UN Climate Panel Stalwarts', by E. Masood.

Nicholls, N. (1996), An Incriminating Fingerprint. *Nature*, vol. 382, 4 July, pp. 27-28.

Nitze, W. A. (1994), A Failure of Presidential Leadership. In I. M. Mintzer and J. A. Leonard (eds.), *Negotiating Climate Change: The Inside Story of the Rio Convention*. Cambridge: Cambridge University Press, pp. 187-201.

New Scientist, 26 August, 1989: 'Soviet Climatologist Predicts Greenhouse 'Paradise'', by J. Miller and F. Pearce.

New Scientist Comment, 8 September, 1990: 'A Climate of Reason'.

New Scientist 11 June, 1994: 'Frankenstein Syndrome Hits Climate Treaty'.

New Scientist, 20/27 December, 1997: 'Dirty Dealings. A coup at Kyoto allows the US to buy the right to pollute', by F. Pearce.

Nordquist, M. H. (1985), *United Nations Convention on the Law of the Sea, 1982: A Commentary*. Vol. 1, Center for Oceans Law and Policy. Dordrecht: Martinus Nijhoff Publishers.

Olson, M. (1965/1971), *The Logic of Collective Action: Public Goods and the Theory of Groups*. Cambridge, Massachusetts: Harvard University Press.

O'Riordan, T. (1997), Betrayers of a Global Truth. Book Review. *Nature*, vol. 389, 16 October, pp. 685-686.

Parson, E. A. (1991). *Protecting the Ozone Layer: The Evolution and Impact of International Institutions*. Cambridge, MA: Kennedy School of Government, Harvard University, CSIA Discussion Paper 92-02.

Parson, E. A. (1993), Protecting the Ozone Layer, in P. M. Haas, R. O. Keohane and M. A. Levy (eds.), *Institutions for the Earth: Sources of Effective International Environmental Protection*, Cambridge, Massachusetts: The MIT Press; pp. 27-75.

Paterson, M. (1992). 'Global Warming'. In C. Thomas (ed.),*The Environment in International Relations*, London: Royal Institute of International Affairs.

Pearce, D. W. et al. (1995). 'The Social Costs of Climate Change: Greenhouse Damage and the Benefits of Control". In J. B. Bruce, H. Lee, and E. F. Haites (eds.), *Climate Change 1995: Economic and Social Dimensions of Climate Change*. Contribution of Working Group III to the Second Assessment Report of the Intergovernmental Panel on Climate Change. Cambridge: Cambridge University Press.

Popper, K. (1963), Science: Conjectures and Refutations. In *Conjectures and Refutations*, London: Routledge and Kegan Paul, pp. 33-59.

Popper, K. (1968), *The Logic of Discovery*. 2nd Edition, New York: Harper and Row.

Popper, K. (1970), Normal Science and its Dangers. In I. Lakatos and A. Musgrave (eds.), *Criticism and the Growth of Knowledge*. Cambridge: Cambridge University Press, pp. 51-59.

Powell, W. W., and P. J. DiMaggio (eds.), (1991), *The New Institutionalism in Organizational Analysis*. Chicago and London: The University of Chicago Press.

Ragin, C. C. (1987), *The Comparative Method: Moving Beyond Qualtitative and Quantitative Strategies*. Berkeley and Los Angeles: University of California Press; chapter 2.

Ramanathan, V. (1988), The Radiative and Climatic Consequences of the Changing Atmospheric Composition of Trace Gases, in F. S. Rowland and I. S. A. Isaksen (eds.), *The Changing Atmosphere*, Chichester: John Wiley & Sons.

Ringius, L. (1997), *Differentiation, Leaders and Fairness: Negotiating Climate Commitments in the European Community*. CICERO, Report 1997: 8.

Rodhe, H., R. Charlson and E. Crawford (1997), Svante Arrhenius and the Greenhouse Effect, *AMBIO*, vol. 26, no.1, pp. 2-5.

Ruggie, J. G. (1975), International Responses to Technology: Concepts and Trends, *International Organization*, vol. 29, no.3, pp. 557-583.

Sanger, C. (1987), *Ordering the Oceans: The Making of the Law of the Sea*. Toronto: University of Toronto Press.

Santer, B. D., T. M. L. Wigley, T. P. Barnett, E. Anyamba (1995), Detection of Climate Change and Attribution of Causes. In J. T. Houghton et al., *Climate Change 1995: The Science of Climate Change*. Cambridge: Cambridge University Press, pp. 407-445.

Santer, B. et al., (1996), Response to F. Seitz, *The Wall Street Journal*, 25 June, 1996, reprinted in *Bulletin of the American Meteorological Society*, vol. 77, no. 9, pp. 1963-1964.

Schelling, T. C. (1960/1980), *The Strategy of Conflict*. Cambridge, Massachusetts: Harvard University Press.

Schneider, S. (1991), Three Reports of the Intergovernmental Panel on Climate Change. Review. *Environment*, vol. 33, no.1, pp. 25-30.

Science, vol. 272, 21 June 1996: 'Industry Group Assails Climate Chapter', by P. Weiss.

Science, vol. 277, 26 September, 1997: 'The Right Climate for Assessment', by R. A. Kerr.

Scientific American, April 1991: 'Profile: Political Engineer. John Sununu dominates science policy', by T. Beardsely.

Scientific American, May 1991: 'Profile: Reluctant Revolutionary: Thomas S. Kuhn unleashed 'paradigm' on the world', by J. Horgan.

Scott, W. R. (1981), *Organizations: Rational, Natural and Open Systems*. New Jersey: Prentice-Hall.

Sebenius, J. K. (1984), *Negotiating the Law of the Sea*. Cambridge, Massachusetts: Harvard University Press.

Sebenius, J. K. (1992), Challenging Conventional Explanations of International Cooperation: Negotiation Analysis and the Case of Epistemic Communities, *International Organization*, vol. 29, no.1, pp. 323-367.

Seitz, F. (1996), A Major Deception on 'Global Warming', *The Wall Street Journal*, 12 June 1996, reprinted in *Bulletin of the American Meteorological Society*, vol. 77, no. 9, pp. 1962-1963.

Shackley, S. and T. Skodvin. (1995). 'IPCC Gazing and the Interpretative Social Sciences'. *Global Environmental Change*, 5:175-180.

Shackley, S. and B. Wynne (1997), Global Warming Potentials: Ambiguity or Precision as an Aid to Policy? *Climate Research*, vol. 8, pp. 89-106.

Shapin, S. (1993), Mertonian Concessions, Book review, *Science*, vol. 259, 5. February.

Singer, F. (ed.) (1992), *The Greenhouse Debate Continued: An Analysis and Critique of the IPCC Climate Assessment*, The Science and Environmental Policy Project. San Francisco, California: ICS Press.

Skodvin, T., J. S. Fuglestvedt (1997), A Comprehensive Approach to Climate Change: Political and Scientific Considerations. AMBIO, vol. 26, no. 6, pp. 351-358.

Skodvin Hegdal, Tora (1991), 'Struktur og aktør i global forhandlinger: Analyse av forhandlingsprosessen i FNs tredje havrettskonferanse. Hovedoppgave (master thesis), Department of Political Science, University of Oslo.

Skodvin Hegdal, T. (1992), 'Structure' and 'Agent' in Institutional Bargaining: Institutional Design and Political Leadership in the Third United Nations Conference on the Law of the Sea. *Cooperation and Conflict*, vol. 27, no.2: pp. 163-189.

Skodvin, T. (in press), The Ozone Regime, in Andresen et al., *Science in International Environmental Regimes: Between Integrity and Involvement*. Manchester University Press.

Smith, S. J., and T. M. L. Wigley (in press), Climatic Implications of Emission Reductions, *Climatic Change*.

Sundquist, J. L. (1978), Research Brokerage: The Weak Link, in L. E. Lynn (ed.), *Knowledge and Policy: The Uncertain Connection*. Washington D. C.: National Academy of Sciences.

Timberlake, L. (1989), The Role of Scientific Knowledge in Drawing up the Brundtland Report. In S. Andresen and W. Østreng (eds.), *International Resource Management: The Role of Science and Politics*. London and New York: Belhaven Press, pp. 117-124.

Torvanger, A. (1997), Rapport frå 8. møte i Ad Hoc Group on the Berlin Mandate (AGBM). (Report from the 8th meeting of the AGBM – in Norwegian). CICERO, Policy Note 1997:3.

Torvanger, A., J. S. Fuglestvedt, B. Holtsmark and L. O. Næss (1997), *Klimaforskning og klimaforhandlinger – status og utsikter fremover*. CICERO Report 1997:5.

Underdal, A. (1980), *The Politics of International Fisheries Management: The Case of the Northeast Atlantic*. Oslo: Universitetsforlaget.

Underdal, A. (1983), Causes of Negotiation Failure, *European Journal of Political Research*, 11, pp. 183-195.

Underdal, A. (1989), International Cooperation: The Art and Science of 'Political Engineering'. Paper prepared for the Annual Conference of the International Studies Association, London, 28. March 1989.

Underdal A. (1990), Designing Politically Feasible Solutions: Notes on the Political Engineering of International Cooperation. Paper presented at the 9th Triannual Convention of the Nordic Political Science Association, Reykjavik; and at the 86th Annual Meeting of the American Political Science Association, San Francisco, August/September, 1990.

Underdal, A. (1991a), International Cooperation and Political Engineering, in Stuart S. Nagel (ed.), *Global Policy Studies*. London: Macmillan.

Underdal, A. (1991b), Solving Collective Problems: Three Modes of Leadership. In Festschrift to Willy Østreng, *Challenges of a Changing World*, The Fridtjof Nansen Institute.

Underdal, A. (1992), Designing Politically Feasible Solutions, in Raino Malnes and Arild Underdal (eds.), *Rationality and Institutions*. Oslo: Universitetsforlaget.

Underdal, A. (1994), Leadership Theory: Rediscovering the Arts of Management, in I. W. Zartman (ed.), *International Multilateral Negotiation: Approaches to the Management of Complexity*. San Francisco: Jossey-Bass Publishers.

Underdal, A. (1997), *Modelling the International Climate Change Negotiations: A Non-Technical Outline of Model Architecture*. CICERO. Working Paper 1997:8.

Underdal, A. (in press), Comparative Conclusions, chapter 8 in Andresen et. al., *Science in International Environmental Regimes: Between Integrity and Involvement*. Manchester University Press.

Underdal, A. (subm.), One Question, Two Answers. In E. Miles et al., *Explaining Regime Effectiveness: Confronting Theory with Evidence*.

Walton, R. E., and R. B. McKersie (1965/93), *A Behavioral Theory of Labor Negotiations: An Analysis of a Social Interaction System*. Ithaca, New York: ILR Press.

Waltz, K. N. (1979), *Theory of International Politics*. New York: Random House.

Watkins, J. W. N. (1970), "Against 'Normal Science' ". In I. Lakatos and A. Musgrave (eds.), *Criticism and the Growth of Knowledge*. Cambridge: Cambridge University Press, pp. 25-39.

Wettestad, J. (in press), From Common Cuts to Critical Loads: The ECE Convention on Long-range Transboundary Air Pollution (LRTAP), In Andresen et al., *Science in International Environmental Regimes: Between Integrity and Involvement*. Manchester University Press.

Winham, G. R. (1977), 'Negotiation as a Management Process' *World Politics*, vol. XXX, no. 1; pp. 87-114.

WMO/UNEP Intergovernmental Panel on Climate Change. Report of the First Session of the Intergovernmental Panel on Climate Change (IPCC), 9-11 November 1988, Geneva.

WMO/UNEP Intergovernmental Panel on Climate Change. Report of the Second Session of the Intergovernmental Panel on Climate Change (IPCC), 28-30 June 1989, Nairobi.

WMO/UNEP Intergovernmental Panel on Climate Change. Report of the Third Session of the Intergovernmental Panel on Climate Change (IPCC), 5-7 February 1990, Washington D.C.

WMO/UNEP Intergovernmental Panel on Climate Change. Report of the Fourth Session of the Intergovernmental Panel on Climate Change (IPCC), 27-30 August 1990, Sundsvall, Sweden.

WMO/UNEP Intergovernmental Panel on Climate Change. Report of the Fifth Session of the Intergovernmental Panel on Climate Change (IPCC), 13-15 March 1991, Geneva.

WMO/UNEP Intergovernmental Panel on Climate Change. Report of the Sixth Session of the Intergovernmental Panel on Climate Change (IPCC), 29-31 October 1991, Geneva, Switzerland.

WMO/UNEP Intergovernmental Panel on Climate Change. Draft Report to the Seventh Session of the Intergovernmental Panel on Climate Change (IPCC), 10-12 February 1992, Geneva, Switzerland.

WMO/UNEP Intergovernmental Panel on Climate Change. Report of the Eight Session of the Intergovernmental Panel on Climate Change (IPCC), 11-13 November 1992, Harare, Zimbabwe.

WMO/UNEP Intergovernmental Panel on Climate Change. Draft Report to the Ninth Session of the Intergovernmental Panel on Climate Change (IPCC), 29-30 June 1993, Geneva, Switzerland.

WMO/UNEP Intergovernmental Panel on Climate Change. Draft Report to the Tenth Session of the Intergovernmental Panel on Climate Change (IPCC), 10-12 November 1994, Geneva, Switzerland.

WMO/UNEP Intergovernmental Panel on Climate Change. Report of the Eleventh Session of the Intergovernmental Panel on Climate Change (IPCC), 11-15 December 1995, Rome, Italy.

WMO/UNEP Intergovernmental Panel on Climate Change. Report of the Twelfth Session of the Intergovernmental Panel on Climate Change (IPCC), 11-13 September, 1996, Mexico City, Mexico.

WMO/UNEP Intergovernmental Panel on Climate Change. Draft Report to the Fourteenth Session of the Intergovernmental Panel on Climate Change (IPCC), 1-3 October 1998, Vienna, Austria.

WMO (1985). 'Report of the International Conference on the Assessment of the Role of Carbon Dioxide and of Other Greenhouse Gases in Climate Variations and Associated Impacts', in Villach, Austria, 9-15 October 1985. Geneva: WMO – No. 661.

Wynne, B. (1993), Implementation of Greenhouse Gas Reductions in the European Community: Institutional and Cultural Factors. *Global Environmental Change*, vol. 3, no. 1, pp. 101-129.

Yin, R. K. (1984/89), *Case Study Research: Design and Methods.* California: Sage Publications.

Young, O. R. (1975), Strategic Interaction and Bargaining. In O. R. Young (ed.), *Bargaining: Formal Theories of Negotiation.* Urbana, Chicago, London: University of Illinois Press, pp. 3-23.

Young, O. R. (1989), *International Cooperation: Building Regimes for Natural Resources and the Environment,* Ithaca and London: Cornell University Press.

Young, O. R. (1991), Political Leadership and Regime Formation: On the Development of Institutions in the International Society, *International Organization*, vol. 45: pp. 281-309.

Young, O. R. (1994), *International Governance: Protecting the Environment in a Stateless Society.* Ithaca and London: Cornell University Press.

Young, O. R., and G. Osherenko (eds.) (1993), *Polar Politics: Creating International Environmental Regimes.* Ithaca and London: Cornell University Press.

Zartman, I. W. (1975), 'Negotiations: Theory and Reality' *Journal of International Affairs*, vol. 9, no. 1; pp. 69-77.

Zartman, I. W. (ed.) (1978), *The Negotiation Process: Theories and Applications*, London: Sage Publications.

Zartman, I. W. and M. R. Berman (1982), *The Practical Negotiator.* New Haven and London: Yale University Press.

Ziman, J. (1968), *Public Knowledge: The Social Dimension of Science.* Cambridge: Cambridge University Press.

Ziman, J. (1984), *An Introduction to Science Studies: The Philosophical and Social Aspects of Science and Technology.* Cambridge: Cambridge University Press.

Appendix

Personal Communication

Interviews, taking the form of informal conversations, have been carried out with the individuals listed below. The positions referred, are positions at the time of the interview, which is also referred.

Dr. A. Al-Gain, vice-chairman of the IPCC, member of the Saudi Arabian delegation, Rome, 5 December, 1995.
Prof. Bert Bolin, Chairman of the IPCC, Oslo, 25 January 1995; 5 January, 1996; and 29 October, 1997.
Dr. Bruce Callander, Co-ordinator of Working Group I, Bracknell, 6 September, 1994.

Mr. Michael Zammit Cutajar, Chairman of the Climate Convention Secretariat, Oslo, 5 January, 1995.
Dr. Richard Derwent, Lead Author of Working Group I, Brighton, 15 March, 1995.
Sir John Houghton, Chairman of Working Group I, Bracknell, 6 September, 1994.
Dr. Ivar S. A. Isaksen, Lead Author of Working Group I, Oslo, December, 1997.
Dr. Richard Lindzen, Professor, Massachusetts Institute of Technology, Oslo 1 May, 1995.
Miss Katherine Maskell, Scientific Assistant at the Technical Support Unit of Working Group I, Bracknell, 6 September, 1994.
Dr. John Mitchell, Lead Author of Working Group I, Bracknell, 7 September, 1994.
Dr. Catherine Senior, Lead Author of Working Group I, Bracknell, 7 September, 1994.
Dr. Susan Solomon, Lead Author of Working Group I, Brighton, 15 March, 1995.
Dr. Robert Watson, Chairman of Working Group II, Geneva, June 1993.

I have also benefited from continual personal communication with members of the Norwegian delegation to IPCC meetings, particularly,

Mr. Øyvind Christoffersen, Norwegian Ministry of the Environment.
Mr. Harald Dovland, Norwegian Ministry of the Environment.
Mr. Lorents Lorentsen, Norwegian Ministry of Finance.
Mr. Håvard Thoresen, Norwegian Ministry of the Environment.

Observation at IPCC meetings:

Ninth Panel Plenary Session, Geneva, June 1993.

Fourth Plenary Session of Working Group I, Maastricht, September, 1994.

Tenth Panel Plenary Session, Nairobi, November 1994.

Second Plenary Session of (the reorganised) Working Group II, Nairobi, November 1994.

Second Drafting Session of Lead Authors of Working Group I to the Second Assessment Report, Brighton, March 1995.

Fifth Plenary Session of Working Group I, Madrid, November 1995.

Eleventh Panel Plenary Session, Rome, December 1995.

Observation at INC meetings:

Ninth Plenary Session of the Intergovernmental Negotiating Committee for a Framework Convention on Climate Change, Geneva, December 1991.

Index

Advances in Global Change Research

1. P. Martens and J. Rotmans (eds.): *Climate Change: An Integrated Perspective.* 1999 ISBN 0-7923-5996-8
2. A. Gillespie and W.C.G. Burns (eds.): *Climate Change in the South Pacific: Impacts and Responses in Australia, New Zealand, and Small Island States.* 2000 ISBN 0-7923-6077-X
3. J.L. Innes, M. Beniston and M.M. Verstraete (eds.): *Biomass Burning and Its Inter-Relationships with the Climate Systems.* 2000 ISBN 0-7923-6107-5
4. M.M. Verstraete, M. Menenti and J. Peltoniemi (eds.): *Observing Land from Space: Science, Customers and Technology.* 2000 ISBN 0-7923-6503-8
5. T. Skodvin: *Structure and Agent in the Scientific Diplomacy of Climate Change.* An Empirical Case Study of Science-Policy Interaction in the Intergovernmental Panel on Climate Change. 2000 ISBN 0-7923-6637-9

KLUWER ACADEMIC PUBLISHERS – DORDRECHT / BOSTON / LONDON

MIX
Papier aus verantwortungsvollen Quellen
Paper from responsible sources
FSC® C105338

If you have any concerns about our products,
you can contact us on
ProductSafety@springernature.com

In case Publisher is established outside the EU,
the EU authorized representative is:
Springer Nature Customer Service Center GmbH
Europaplatz 3, 69115 Heidelberg, Germany

Printed by Libri Plureos GmbH
in Hamburg, Germany